Engaged Encounters

ENGAGED ENCOUNTERS

Thinking about Forces, Fields and Friendships with Monique Nuijten

edited by:

Elisabet Dueholm Rasch
Oscar Salemink
Bram Büscher
Michiel Köhne

Wageningen Academic
Publishers

EAN: 9789086863563
e-EAN: 9789086869091
ISBN: 978-90-8686-356-3
e-ISBN: 978-90-8686-909-1
DOI: 10.3920/978-90-8686-909-1

About the cover
The cover is inspired by a picture from Monique Nuijten's photo archive. It was, most probably, taken somewhere between 1991 and 1994 by Pieter de Vries, in La Canoa (Mexico) during Monique's fieldwork. It features a farmer, who was one of the key informants for Monique at that time, with his wife.

First published, 2020

TABLE OF CONTENTS

PART 3: PERSONAL ENCOUNTERS 165

INTRODUCTION: ENCOUNTERING AND ENGAGING MONIQUE NUIJTEN

Bram Büscher, Michiel Köhne, Elisabet Dueholm Rasch and Oscar Salemink

Engaged encounters

Engaged encounters define the life and work of Monique Nuijten, the eminent scholar to whom this festschrift is dedicated. From her academic home base in Wageningen, the Netherlands to the various countries in Latin America and Europe where she conducted field research, Nuijten embodies the ideal of a politically engaged ethnographer who is able to combine astute empirical observations with innovative conceptual interventions. For many colleagues around the world she also exemplifies an engaged collegiality that is nothing short of exemplary. Those who had the privilege to work closely with Monique Nuijten can all testify to her innate ability to combine the political and collegial with the personal in a way that often led to warm friendships.

The contributions to this book showcase these three principal dimensions of engaged encounters, which in Nuijten's work emerge as three interconnected but distinct foci. On the most abstract level, 'engaged encounters' appear as a *way of being in and thinking about the world*. Being in the world is often understood as a state (of mind, of being). This state is dynamic but at the same time often experienced as static, as something that holds a particular permanence or essence. This classic idea of static being and thinking holds little resonance for Monique Nuijten. While she would acknowledge certain permanences, Nuijten always stresses the relational in her work whereby people on different levels and within different institutions are, first and foremost, embodiments of historical, contemporary and future encounters. Persons exist in and through encounters across time and space, and these are necessarily engaged in that they involve historical, political and personal stakes. These encounters shape the institutions and organisations that people create and inhabit, as well as their everyday practices.

This distinction between institutions and organisations and how these are enacted in daily life is another key focus of Monique Nuijten's work. This leads us to the second way to approach 'engaged encounter', namely as a *political way of interacting*. Engaged encounters are always political encounters. Because actors have fundamental stakes in the encounters that make up their world, they necessarily partake in these politically. Their interests conjoin with others and with historical and contemporary contexts to become force fields. In these force fields that constitute their life-worlds along intertwined structural and personal dimensions, actors engage affectively as well as strategically: thinking through and about the forces that impact their lives, which simultaneously infuses their actions. Engaged encounters thus become political in two ways: as structural forms of organisation and as modes of agency. Together they entwine the regularities and irregularities, the forms of domination and facilitation, and the expectations and apathies that characterise real-world interactions.

Third, 'engaged encounters' concern *self-definition and self-reflection*, in relation to two central Nuijtenian conceptualisations of the world: the world as a whole, the one we all create together and in which we encounter each other; and the life-world of the persons that embody and enact these broader connections. As people define and reflect on the world around them and how their own life-world fits within this broader environment, they themselves come to embody engaged encounters. This dimension of engaged encounters is therefore also a question of identity and culture, of the complex entanglements of persons and collectives. In other words, the reflective self sits uneasily between, but also connects, the two worlds as an engaged encounter in and of itself, and necessarily ends up stretching the personal to also encompass the collective.

Monique Nuijten's life and work express these three dimensions of 'engaged encounters' – the theoretical, the collegial and the personal – that structure the three parts of this festschrift. As many of the ensuing contributions emphasise, theory for Nuijten is intensely personal, as is her scholarship more generally. This we refer to as 'the collegial', not just to underline her close way of working with academic colleagues, but especially to emphasise the meaning of collegiality as reciprocal and mutual. Nuijten believes that the people that we live, interact and engage with during fieldwork should be approached as colleagues with whom the researcher develops reciprocal bonds of mutuality, rather than respondents or – more technocratically – 'research subjects'. This is not to say that these bonds are equal or equanimous. They are, again and always, *political* – but bonds nonetheless; bonds between colleagues and as such also between theory and the personal. In practice, however, and perhaps precisely because of Nuijten's insistence on the political, she has been able to develop forms of mutuality and reciprocity with colleagues and people encountered during fieldwork that often turned into lasting friendships, as the forty contributions to this festschrift, all in their own way, attest.

The three parts of the book follow the three approaches to engaged encounters and relatedly situate Monique Nuijten's work in between forces, fields and friendships. In the next section we briefly examine the insights, questions and themes to which this leads.

Theoretical, collegial and personal encounters

The 40 chapters in this book are loosely organised in three parts focused on theoretical, collegial and personal encounters – 'loosely', because the distinctions between these are oftentimes spurious, and many contributions in fact show the deep connections between them in Monique Nuijten's scholarship. At the same time and informed by the specific and unique encounters they have had with Nuijten's work or her personally, the authors emphasise different aspects of her scholarship that we aimed to capture under the three headings of theoretical, collegial and personal encounters. So, what are some of the main insights, questions or themes that come to the fore in the three parts?

The first and most voluminous part on *theoretical encounters* contains eighteen contributions, which range from reflections on Monique Nuijten's contributions to our understanding of organising practices and legal anthropology to engagements with her more recent work on 'the urban'; via reflections on the state, development and hope, and force fields. Quite a few authors emphasise and engage with Nuijten's two most influential concepts, namely development as a 'hope-generating machine' and that of 'force fields'. One key insight that emerges is that these two concepts are most fruitfully seen together: namely that it is precisely within environments infused with politics and power that processes of development come to define not just more standard ideas of 'progress' or growth but that for many people they translate into bureaucratic and institutional forms that offer, and themselves come to depend on the idea of, hope. Another insight is that these forms connect and transcend the rural and the urban in ways that fundamentally change how we think about those categories.

The second part comprises eleven chapters assembled under the rubric of *collegial encounters*. The background insight of this part derives from the fact that universities are – unfortunately but not surprisingly – replete with troublesome stories about fierce competition, arrogance, indifference, gossip, envy and even back-stabbing or worse. Against these and other deeply problematic aspects of modern-day academia, radically different, engaged forms of collegiality and scholarship, such as those practiced by Nuijten, stand out. What is especially apparent from the chapters is that Nuijten's way of being in the world is very much in line with her ethnography and theorisation. The image that emerges from these chapters is of a scholar who can be razor sharp and confrontational in her observations and comments, while remaining humble, warm and kind. This, as many of us know, is rare: combining and embodying reciprocity, mutuality and generosity in ways that make others feel at ease, while not shying away from frictions or the troublesome aspects of the force field of academia that needs highlighting. In that sense, Nuijten herself quite literally embodies 'hope-generation': defying bureaucratic structures and impositions in a time when higher education is marked by intense neoliberalisation and the troubled work-floor tensions and pressures to which this has led over the last 30 years.

The third part contains eleven encounters where the personal, the political and the scholarly merge most concretely in Monique Nuijten-as-person. The authors draw on their personal encounters with, and recollections of, Monique to draw a multi-faceted but consistent picture of her as an inspiring colleague, mentor, supervisor, friend and – in one case – life partner, whose embodied scholarship is an integrated part of her life. Despite her own eschewing of the epithet of 'feminist', Monique emerges as a female role model for many younger colleagues. The chapters also show that as Monique herself navigated and combined the various intellectual, societal, political and personal roles in her daily life she was open about the frictions that come with them. This quality of bringing out tensions and frictions in a way that makes acknowledging and dealing with them possible and often easier enabled Monique to consistently lighten up the force fields in which she moved while making a lasting mark on her friends and colleagues as well as on the knowledge they produce.

Clarifying differences, crossing boundaries

As is clear from the foregoing, the three dimensions of engaged encounters, and the book parts that arise from these, morph into each other and define each other. If they can be distinguished at all, they certainly cannot be separated. Yet what is remarkable about Monique Nuijten's work is that this does not lead to fuzzy analyses. In crossing boundaries, Nuijten clarifies differences, makes them political and places them in broader frameworks of power. Through key concepts and interventions, such as force fields and development as a hope-generating machine, Monique Nuijten clarifies for many of us how engaged encounters may seem fuzzy and crystal clear at the same time: fuzzy because of how they connect different areas of concern; and clear by showing that the stakes of variegated encounters lead some actors – research participants, but also colleagues, students and friends – to benefit and others to be marginalised. This, then, is the final insight we want to highlight for the reader who moves through the ensuing parts and chapters: in political force fields, the relational as it develops through theoretical, collegial and personal encounters cannot be an analytical end in itself. Rather: in our analyses, it must illuminate the stakes involved in and through these encounters to highlight, open up and resist forms of oppression and domination. This and many other insights from Nuijten's life and work highlighted by the contributors have enriched development sociology in ways that – is our staunch hope! – will continue to encounters by generations of scholars to come.

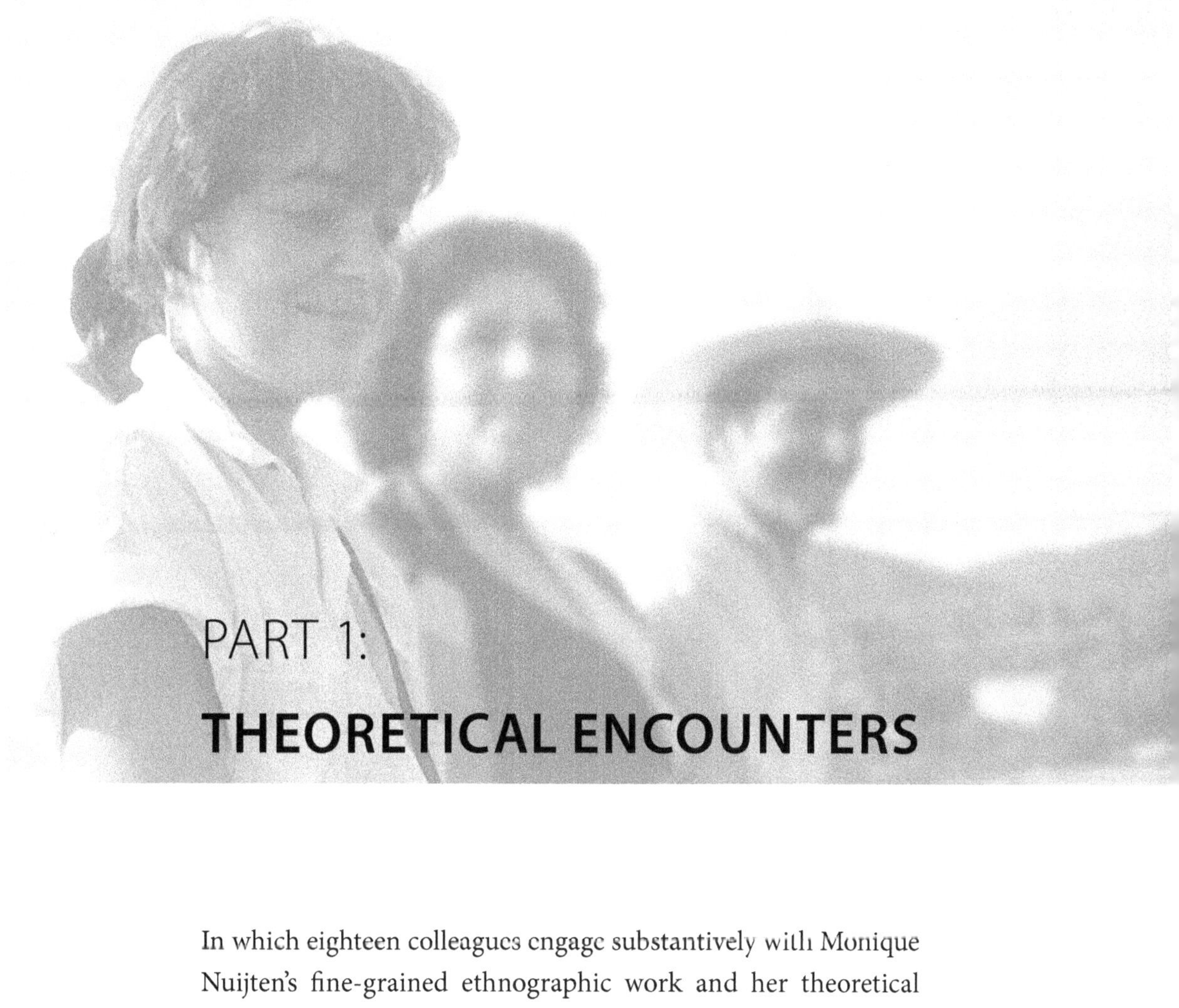

PART 1:
THEORETICAL ENCOUNTERS

In which eighteen colleagues engage substantively with Monique Nuijten's fine-grained ethnographic work and her theoretical interventions and contributions. These are inspired by her unique personal and political habitus, and have in turn inspired others' scholarly, personal and political engagements in various ways.

1. CAPTURING LAW: MONIQUE NUIJTEN'S CONTRIBUTION TO THE STUDY OF PLURAL LEGAL ORDERING

Keebet von Benda-Beckmann

What characterises Monique Nuijten's work as I have learned to appreciate over the years – is her keen eye for discerning fluidity and flexibility in social interaction, layered understandings, multiple positions of members of transnational communities, and active attempts of people to bring coherence into their life while coping with the possibilities and adversities in their particular situation. She draws on a broad range of theoretical approaches and energetically engages in debates with a diverse set of authors that have inspired her work. Sceptical of grand theories that highlight powerful actors, she prefers to divert her attention to people from all walks of life, who operate within a range of social settings, some with a certain degree of power, others without much power, some who have been successful and others that have not been able to fulfil their dreams.

Law has not been at the forefront of her theoretical work, though she has always been aware of the extent to which legal issues are vital to her empirical work. She is cognisant of the fact that the people she studied in Mexico, in the US, Peru, and Brazil were heavily influenced by the laws of the respective states. One might say that her work tangentially intersects with the anthropology of law and the study of legal pluralism. She has engaged with all her energy in discussions about law and legal pluralism that took place in the 'theory evenings' at our home in Wageningen, where my husband Franz von Benda-Beckmann taught anthropology of law at Wageningen University, and when we visited her in Mexico during her field research that eventually resulted in her book *Power, Community and the State* (Nuijten, 2003).

At the conferences of the Commission on Legal Pluralism and at the Max Planck Institute for Social Anthropology in Halle, where she attended a series of conferences on core issues in the anthropology of law (property, governance, moving law and migration, and space), her contributions centred on questions of legal ordering. Franz and I immensely enjoyed these discussions with her. We tried to convince her that she was doing important work that contributed a vital dimension to the study of legal pluralism. She, however, has remained uncomfortable with that term and prefers other concepts to characterise her observations. I believe the main reason for her discomfort is that she associates the terms law and legal pluralism with something solid, clearly demarcated, stable, and imposed. This may have been true for some studies on legal pluralism, but certainly not for the entire field of the anthropology of law. While most have worked using some variant of actor-structure theories with a keen eye for the interaction of actors, her focus has definitely been on the actor side, while for us 'structure' has been equally important. At times, it seemed as if she rejected the notion of structure altogether, but it would reemerge in her characterisation of how practices, procedures, norms, and values were organised. Our discussions with her sharpened my understanding of law as an umbrella concept, which, as I see it, includes the issues she is interested in. Despite these terminological – and probably substantive differences, it is clear that her rigorous focus on norms and values used and generated in social interaction has enabled her to make important contributions to the anthropology of law (Von Benda-Beckmann and Turner, 2020: 86, 131).

Her insight in the social significance of law is crucial in that it concerns the way in which state law affects social relations in general, even though her study more poignantly focused on Mexican *ejido*s that were established following the land reforms introduced between 1917 and 1992. Probing into the history of property relations in an *ejido*, Monique shows how the *ejido* became the core of a discursive framing of property relations by which landless residents of the village remained excluded from official membership of the *ejido*. The *ejido* is thus captured and used to resist subsequent legislation that would allow non-member residents more equal property rights to the commons (Nuijten, 1997, 2003, 2006). The general point of these studies is that state law, or parts thereof, enacted at a certain time in history, may be captured by the local population over time. They may appropriate state regulations as their own and remodel it to adapt it to their own 'organizing practices' as they befit their needs and understandings. These sets of norms and procedures then serve as guidelines in negotiations, though the outcome may be quite unpredictable. As local norms, these sets of state regulation often linger and continue to be used even after the government enacts new laws. Thus, in a recursive process, state law may transform into a local normative order, though Monique would be hesitant to use the term law for this, and would prefer to say that they become embedded in local organizing practices in which people order their life and social relations. Such capturing of state law and its lingering in locally adapted versions far beyond its expiry date was a new insight for the anthropology of law.

A second set of issues to which Monique has made important contributions concerns the way norms change with migration (Nuijten, 2005). Building on the seminal publication by Sally Falk Moore (1973) on the semi-autonomous social field, and on Nina Glick Schiller's work on transnational social fields (2005), she has pointed to the dynamics by which the internal norms and values of a transnational community of Mexicans having migrated to the USA are reframed. How these norms are reframed depends on the legal status individual Mexican members of such a community have under American law, varying from illegal to a fully legal status and everything in between. This has implications not only for how migrants operate in the US, but also 'back home' in Mexico (though for youngsters it may not be home at all). Returning migrants discover that differential reframing is happening in their transnational community as well. Whether temporarily returning as successful migrants or permanently because migration had been unsuccessful, they all have to deal with and participate in shaping the diverging norms and values that exist locally. The dynamics within transnational communities turn out to be vital for the plural legal orders at both ends of migration movements.

In short, what makes Monique Nuijten's work unique in the field of legal pluralism is her capacity to engage with the lives of individual persons in a diverse range of bureaucratic and other social settings, and her consistent focus on people's interaction along various hierarchies and capabilities underpinned by varying levels of power and access to resources. This had enabled her to access and understand layered interpretations of norms and capture the dynamic and fluid normative patterns of legal ordering as they emerge over time.

Keebet von Benda-Beckmann / Professor Emeritus, Martin Luther University Halle, Wittenberg / associate Max Planck Institute for Social Anthropology, Halle, Saale, Germany / kbenda@eth.mpg.de

References

Glick Schiller, N., 2005. Transborder citizenship: an outcome of legal pluralism within transnational social fields. In: Von Benda-Beckmann, F., Von Benda-Beckmann, K. and Griffiths, A. (eds) Mobile people, mobile law. Expanding legal relations in a contracting world. Aldershot and Burlington: Ashgate, pp. 27-50.

Moore, S.F., 1973. Law and social change: the semi-autonomous social field as an appropriate subject of study. Law & Society Review 7(4): 719-746.

Nuijten, M., 1997. Privatization of the ejido in Mexico: the interplay between legal change and local practice. In: Von Benda-Beckmann, F., Von Benda-Beckmann, K. and Hoekema, A.J. (eds) Natural resources, environment and legal pluralism. The Hague: Martinus Nijhoff, pp. 72-104.

Nuijten, M., 2003. Power, community and the state: the political anthropology of organization in Mexico. London, Pluto Press.

Nuijten, M., 2005. Transnational migration and the re-framing of normative values. In: Von Benda-Beckmann, F., Von Benda-Beckmann, K. and Griffiths, A. (eds) Mobile people, mobile law. Expanding legal relations in a contracting world. Aldershot and Burlington: Ashgate, pp. 51-68.

Nuijten, M. and Lorenzo, D., 2006. Moving borders and invisible boundaries: a force field approach to property relations in the commons of a Mexican ejido. In: Von Benda-Beckmann, F., Von Benda-Beckmann, K. and Wiber, M.G. (eds) Changing properties of property. New York and Oxford: Berghahn Books, pp. 218-242.

Von Benda-Beckmann, K. and Turner, B., 2020. Anthropological roots of global legal pluralism. In: Berman, P.S. (ed.) The Oxford handbook of global legal pluralism. Oxford: Oxford University Press, pp. 67-141.

2. ORGANIZING PRACTICES, PROPERTY, AND SOCIAL MOVEMENTS: AN IMPRESSIVE CONTRIBUTION TO SOCIOLOGY OF DEVELOPMENT, POLITICAL AND LEGAL ANTHROPOLOGY

Dik Roth

I still remember that, it must have been in 1992 or 1993, I first heard Monique Nuijten present her work during a seminar organised at the Vrije Universiteit Amsterdam, bringing together critical academics, many of whom were related to the EIDOS (European Inter-University Development Opportunities Study-Group) network, and development practitioners. I had just returned from Indonesia, where I had been an adviser of a conflict-prone and contested land reform programme. It was this seminar, and more specifically Monique's work on the land conflicts around the Mexican *ejido*, that made a great impression on me. The seminar connected me with critical debates on bureaucracies, development interventions, and actor oriented sociology, discussions I had missed during my Indonesia years. Monique's work, more specifically, engaged more directly with issues of interventions in land, land reforms, and property conflicts. These were exactly the kind of themes and issues that I was trying to make sense of in relation to my Indonesian experiences. Wageningen University, with its sociological focus on rural and agrarian change through actor-oriented sociology and legal anthropology, was the place where many of these debates were developing. It was also the place where Monique had been studying and had then started doing her PhD.

Some years later, I joined Wageningen University to do my PhD with the Agrarian Law Group. Monique and I were members of different chair groups at the Social Sciences department. As teachers we were never involved in the same courses, sometimes we met at student seminars or other events. It was at the conferences abroad that we interacted most. Sharing scientific interests in ethnography of development interventions, anthropology of law and property, we participated in similar networks. We joined two EIDOS conferences in London and the legal pluralism conference at Chiang Mai together. We also participated together in several meetings of the Rural Property Network in the early 2000s. I always enjoyed these interactions very much. The various meetings provided an opportunity to get to know Monique's work through her always interesting presentations, and Monique as a Wageningen colleague through our conversations while travelling to London or staying in Chiang May. More recently we became colleagues in the newly formed Sociology of Development and Change Group, and shared an office in the last few years. Our current group finally combines several of the closely related themes that have long remained artificially separated in the organisational structure of the university bureaucracy, but have —in combination!— been at the heart of Monique's work from the early 1990s until now, as I will show below.

It is impossible in this short contribution to do full justice to the impressive contributions Monique has made to various fields, from development studies to anthropology of law and activism. In this contribution I will shortly discuss what are, in my opinion, three important dimensions of, or maybe also periods in, Monique's work. In the first, early period, we can see a critical engagement with development policies and interventions against the background of the role of the state and its bureaucracies, approached from an actor-oriented perspective. In the second period a further engagement with issues and questions of property stands out, grounded theoretically in both actor-oriented sociology, legal anthropology and critical theories of the state. In more recent work, I see an 'urban turn' and new engagement with research in Europe (Spain). While Monique continues engaging with the earlier debates that have inspired her work through her career, the shift towards urban-based work also inspired a move towards engagement with social movements and marginalised groups in urban settings.

In her earlier work (e.g. Nuijten 1992), Monique already shows a distinct critical scientific interest in organizing processes and practices of development interventions. This was part of a broader criticism of both mainstream formal bureaucratic approaches and ideologically inspired 'empowerment' approaches that were increasingly popular in development policies but, like the former, failed to seriously engage with everyday practices of organizing in specific contexts. In both, relations between formal organisation and behaviour are assumed on the basis of normative organisational formats rather than analysed. This normative content also generated Monique's deeper interest in the 'legal' part of the story. Her work on the *ejido* basically involved a comparison of the formal legal model and organisational procedures of the *ejido* with its local organizing practices, without normatively (and usually negatively) labelling the latter in terms of deviation, informality or otherwise. These ideas, themes and theorisations were already prominently present here, but further developed of course in her PhD thesis (Nuijten, 1998: 2003). Her research on the *ejido* also meant paying critical attention to the role of the state in land laws and reforms, and in attempts to control rural people and property, the meanings given locally to organisational resources and practices, and questions of legitimacy of state ruling and political-administrative control. In this period, Monique's work reflects both the influence of the theoretical perspectives and approaches developing in the prevailing academic 'force fields' at the Social Sciences Department of Wageningen University in the late 1980s and 1990s, as well as the great creativity with which Monique developed her own themes, foci and approaches in this academic context and applied them to the *ejido*.

Building on her PhD research and with ongoing scientific interest in the *ejido*, at a later stage in her work, Monique wrote a number of interesting articles and book chapters on the *ejido* as a field of contested property. This gave her (and her co-authors) the possibility to engage with legal anthropological theories of property, to discuss the *ejido* as a 'commons', and further explore interventions in property in a wider context of transformation of relations of production, legal and other institutions and government policies and programs. Using the practice-force field approach to property relations, Nuijten and Lorenzo (2006), for instance, discuss the changing access and use practices in the *ejido* and the discursive framings of land claims in a social context in which the meaning of the *ejido* (and thus of 'property') goes well beyond the economic value of land. They show how the uses, forms of access and meanings of common land were historically changing in a changing socio-economic context. Property relations around *ejido* land are deeply embedded in the power relations between landowners and landless residents, between those whose social position is secure and gives them the power to exclude, and those who are dependent on others for a permission to be included. This is all placed in a wider context of ongoing influence of the historical and ideological meanings of the *ejido* and of often contradictory government policies (see also Nuijten, 2004). In a later contribution (Nuijten and Lorenzo, 2009), the same authors expand this property focus in a case study of property relations in a Peruvian peasant community. In this article they explore the basis of legitimacy of governmental technologies of governance as expressed in approaches to land based on rule-making, registration and quantification practices. These practices both strengthen the legitimacy of government and provide new options for peasants to hold the government accountable. As the authors conclude: 'Registration and quantification can never be complete or accurate, but that is not the same as being completely inaccurate and false. Democracy is always a "limited democracy" (2009: 102).'

In Monique's most recent work a gradual 'urban turn' becomes visible. Her ongoing engagement with social movements, marginalised groups and grassroots political agency is now, more than before, turned to urban settings and 'force fields' (see e.g. Bode and Bakker, 2018; Koster and Nuijten, 2011; Nuijten, 2015). In the latter, using a Mouffian perspective on 'the political' (2015: 476), Monique's focus lies on those people 'that have a contested relationship with the existing political order, namely 'people in the margins' and 'activists' (Nuijten, 2015: 477)'. An example is her critical engagement with the problem of massive house evictions in Spain, through research on the *Plataforma de Afectados por la Hipoteca*; PAH, Platform of Mortgage Victims). The normative values inspiring grassroots activist groups like PAH may differ widely from those of the neoliberal capitalist state and its rules and policies, which they regard as incompetent and serving the interests of the elites. These frictions and contestations can be expressed in many ways, such as street art forms, music and dance (a topic that was explored by one of Monique's students and discussed in an interesting article; see Van Leerzem *et al.*, 2015). As the case of grassroots support to Spanish mortgage victims (Nuijten, 2015) shows, another option for protest is the open confrontation with, and contestation of, the state and its rules, as well as the banking rules and practices that have caused the financial problems of mortgage victims. Such more direct contestations can involve collective resistance, house occupations and practices like making demands on the government and publicly condemning and approaching key political and economic actors. As explained by Monique, such forms of engaged ethnography of grassroots political agency with activists, victims of mortgages and other actors require specific research methods, including open dialogues in which alternative political futures can be jointly explored. Thus, we can see a more direct engagement with protesting groups in an urban environment. In this new urban focus, earlier foci continue to play an important role, such as the role of law and normative pluralism (Nuijten, 2015) and of participation and inclusion in project intervention practices (for urban 'slum upgrading') (Koster and Nuijten, 2011).

Monique's research agenda and academic work shows a consistent development characterised by both continuities and changes. Theoretical orientations show a high degree of continuity, with actor-oriented sociology, critical development studies, legal anthropology (of property) and political anthropology as important sources of inspiration. In more recent work, engaged ethnography, applied to research projects in new, mainly urban settings, has become more prominent in her work. This has become a further enrichment of an already impressive contribution to various fields, mainly sociology of development, political anthropology and legal anthropology. These are, and will remain, core domains in the Sociology of Development and Change Group, to which Monique has contributed so much until today.

Dik Roth / Sociology of Development and Change Group, Wageningen University, the Netherlands / dik.roth@wur.nl

References

Bode Bakker, M. and Nuijten, M., 2018. When breaking you make your soul dance. Utopian aspirations and subjective transformation in breakdance. Identities 25(2): 210-227.

Koster, M. and Nuijten, M., 2011. From preamble to post-project frustrations: the shaping of a slum upgrading project in Recife, Brazil. Antipode 44(1): 175-196.

Nuijten, M., 1992. Local organization as organizing practices. rethinking rural institutions. In: Long, N. and Long, A. (eds) Battlefields of knowledge. The interlocking of theory and practice in social research and development. Routledge, London and New York, pp. 189-207.

Nuijten, M., 1998. In the name of the land. Organization, transnationalism, and the culture of the state in a Mexican Ejido. PhD thesis, Wageningen University, Wageningen, the Netherlands.

Nuijten, M., 2004. Peasant 'participation', rural property and the state in western Mexico. Journal of Peasant Studies 31(2): 181-209.

Nuijten, M.C.M. and Lorenzo, D., 2006. Moving borders and invisible boundaries: a force field approach to property relations in the commons of a Mexican ejido. In: Von Benda Beckmann, F., Von Benda Beckman, K. and Wiber, M. (eds) Changing properties of property. New York and Oxford: Berghahn Books, pp. 218-242.

Nuijten, M., 2015. Political activism and the PAH (Platform of Mortgage Victims) in Spain: the significance of Franz von Benda-Beckmann's work for the study of power, political agency and legal pluralism at the grassroots. Journal of Legal Pluralism and Unofficial Law 47(3). 476-492.

Nuijten, M. and Lorenzo, D., 2009. Ruling by record: the meaning of rights, rules and registration in an Andean comunidad. Development and Change 40(1): 81-103.

Van Leerzem, L., Nuijten, M. and De Vries, P., 2015. Being negro in Recife Brazil: a political aesthetics approach. Ethnos 81(5): 865-887.

3. A CONVERSATION WITH MONIQUE NUIJTEN

Gavin Smith

I would like to engage with Monique's work by conducting a kind of conversation with her, or with some of the ways I understand her to be thinking as I read her work. We come from different generations. When my anthropology might have been bitten by the 'development' temptation both anthropology and development meant different things from the early 2000s when Monique was completing her doctorate. I began my career as an anthropologist at a time when Eric Wolf and his anthropology colleagues were organizing the Teach-ins in response to the US war on (sic) Southeast Asia. In fact, Wolf's *Peasant Wars of the 20th Century* played a significant part in what I chose to study as an anthropologist. Conventional anthropology was under attack but so was 'development' associated as it was with Rostow's (1960) *The Stages of economic growth* and – still more insidious – David McClelland's (1967) *The achieving society*. While in Europe anthropology in the sixties was relatively unaffected by the temptations of 'changing the world' through development programmes, the same was not the case for North America where via 'development,' anthropology could not detach itself from the US's Cold War agenda in the global south. Some were for it (both the development and the agenda). My bunch were against it. But whoever you were, you couldn't leave it alone. As Eldridge Cleaver famously said, 'There is no more neutrality in the world. You either have to be part of the solution, or you're going to be part of the problem.'

I cannot of course, speak for Monique. But I can speculate. Her supervisor was Norman Long who had been doing fieldwork in Peru at the same time as I was. His research project was a cross-over sociological-anthropological project with Bryan Roberts, both from Manchester University home of what we anthropologists call 'the Manchester School'. Let me risk surmising that what Norman picked up from that kind of anthropology might have rubbed off on Monique, especially in her use of grounded theory as a note below. But I think it's fair to say that by the time Monique began her research the whole enterprise of 'development' had perforce gone through a sea change. Obviously this was not unconnected to the events of 1989, but long before that development as 'stewardship' (Cowen and Shenton, 1996) had been overtaken by the entry of 'participation' into the vocabulary of development.

So, between 1975 when I submitted my thesis, and the early 2000s when Monique submitted hers, anthropology-sociology and development studies had become different animals. I want however to reflect on three themes, issues, imaginaries or what-have-you, in my work and hers whose intertwining are especially interesting. They come under the headings: *forms of praxis*; *theory and method*; and *sources of agency*.

Monique has written about development as a 'hope generating machine' which I take to mean that projects employ hope to put fuel in the tank because, unlike stewardship, this kind of development requires the activation of the people. This, I would say, is the way developmentalists understand *praxis* and they start designing their methods right from the outset by taking this into account. It is hard to believe these days, but the fact is that *praxis* as revolutionary struggle had a similar relationship to hope. The works of those who published in the sixties like Edward Thompson, Eric Wolf, Eric Hobsbawm, or Stuart Hall cannot possibly be understood except as contributions to the politics of hope – toward socialist revolution… and the kind of *praxis* that might require. That hope is of course dead. People write differently about revolution these days. The anthropologist David Scott (2014) for example speaks of it in terms of tragedy. But perhaps the wheels are falling off 'development' as a hope generating machine too.

In fact, the tension that arose in the sixties between the cold war developmentalists and those more inclined towards a more revolutionary kind of intervention reveals I think a contradiction deep within developmental agendas across the eras. What exactly *is* the praxis a development practitioner wants to invoke? Just how revolutionary does such a person want their project to be? Or is development in truth, that is as we have always seen it practised, from beginning to end *the* anti-revolutionary project? In this sense, contrary to those who speak of the perpetual failure of development projects to achieve their goals, insofar as socialist revolutions are a thing of the past, perhaps we could say that development projects taken as a whole have succeeded: Walt Rostow maybe laughing in his grave. In any event I would venture to say that throughout her work Monique has worried away at this kind of dilemma. What is the relationship between the down-to-earth practices of the people and the *praxis* development practitioners want to achieve? How empowering is it? Or to speak more precisely in Monique's terms how do such projects reshape the field forces of power?

And what has given her theoretical interventions their powerful critical edge is her unsurpassable ethnographic engagement, what Michael Burawoy (1991) would call her grounded theory. Reading through her work, you will never find a theoretical assertion that is not powerfully demonstrated by what she tells us of people acting, interacting, speaking, arguing, narrating; and not just where Monique has been doing the fieldwork but across a wider template of scales that she has also subjected to interrogation. Which brings me to my third 'theme' – the question of agency. I would say that this is an issue that has always fascinated Monique and that she has insistently threaded through the varying ways in which forms of power enable agency, restrict it, but crucially actually form it: give agency its form.

Yet, at least for me the term 'agency' raises difficult questions about scale. When we speak of an actor who exercises leverage in social action, what time scale should we be using and what kind of subject are we speaking of (from individual subject to collective subject, an issue of particular concern to Gramsci, for example)? In my case when I became interested in rural rebellion as a means by which people sought to increase their agency, I turned to books like Wolf's (1969) *Peasant Wars* and Hobsbawm's (1959) *Primitive Rebels*. But what I found there was very far from the kind of grounded theory we see in Monique's work. For me what was missing was what a group of rural people, or a similar group of impoverished urban dwellers, have to do so as to work collectively to make an impression on the political field. In the context of the elusive *habitus* of their daily lives how might they achieve some kind of subaltern hegemony (Thomas, 2009, 2018) That seemed to be missing from the books I was reading. They had in common a particular perspective, what today we might call a sort of positionality: a view of different rural movements from the outside looking in. For me catching the dynamics of those relations and the way in which ideas emerged momentarily in the actuality of practice was a priority: a kind of enquiry in which the enquirer tries to 'enter the space of the world the researcher seeks to understand.' (Ortner, 1995: 173) How did such people reign in the fractious inclinations of self-interest to enhance the agency of the whole?

It seems to me that Monique has long been asking a similar question but from a very different, almost an opposite, angle: developmentalists persistently want to induce collectivity, or simply imagine its possibility, but they do so by blinding themselves to the way interactive agency operating in what she calls force fields run through any and all forms of collectivity – movement, institution, church and so on – like the crazy paths of iron finings as power (literally – different kinds of power) is turned on or off. So, what seemed to me to be rather taken for granted by the big writers on rural revolution is something, as I see it, that has formed a central target of Monique's critical theory too. And, as participatory development has become ever more entwined with grass roots movements, so the line between my own concern with the enhancement of agency through collective struggle has become vulnerable to the kind of acid test to which

Monique subjects it through the many instances of her grounded critical enquiries. True of course Monique's pursuit of what I might call 'the truth of action' has changed over the years – since her force fields of power article in 2005, for example, to her study of the participatory procedures of the Brazilian state in Recife in her 2013 article, and still more recently in her interest in 'the secret of law' in 2016 – but the on-the-ground engagement, the fascination with not just what people do but how they think about what they do, and the impressive combination of pursuing her ideals while refusing the siren call of romantic solutions, these have always been the hallmark of her work.

Gavin A. Smith / Professor Emeritus in Anthropology, University of Toronto, Canada / gavin.gav@gmail.com

References

Burawoy, M., 1991. Ethnography unbound: power and resistance in the modern metropolis. Berkeley: University of California Press.

Cowen, M. and Shenton, R., 1996. Doctrines of development. Routledge.

Hobsbawm, E.J., 1959. Primitive rebels: archaic forms of social movements in the 19th and 20th centuries. Beacon Press, Boston.

McClelland, D., 1967. The achieving society. Free Press, Boston.

Ortner, S., 1995. Resistance and the problem of ethnographic refusal. Comparative Studies in Society and History 3(1): 173-193.

Rostow, W., 1960. The stages of economic growth, a non-communist manifesto. Cambridge University Press.

Scott, D., 2014. Omens of adversity: tragedy, time, memory, justice. Duke University Press.

Thomas, P., 2018. Refiguring the subaltern. Political Theory Online 46(6): 861-884. https://doi.org/10.1177/0090591718762720

Thomas, P., 2020. After (post) hegemony. Contemporary Political Theory in press. https://doi.org/10.1057/s41296-020-00409-1

Wolf, E.R., 1969. Peasant wars of the twentieth century. New York: Harper & Row.

4. NAVIGATING NYAMAROPA: THE SIGNIFICANCE OF LOCAL FORMS OF ORGANISATION

Paul Hebinck

Introduction

Monique Nuijten is known for her work and understanding of *local forms of organisation* which she began to explore in the early 1990s in an *ejido* in Mexico (Nuijten, 1998; Nuijten, 2003). She analysed the *organizing practices* of the *ejidatarios* by looking at the significance of the *ejido* in their livelihoods and by studying their relationship with the wider socio-political context. This inevitably enters the debate about state-peasant relationships. The tendency in much of the agrarian studies is to adopt a political economy framework which (re)produces an image of the peasantry (or the *ejidatarios*) as dependent social actors and passive recipients of state services (e.g. markets, knowledge, technologies). Like Norman Long (Long, 2001), she finds these interpretations to assign too little importance to the role of local practices and knowledge in development as these tend to reinforce the image and efficacy of conventional top-down designed policies. In contrast, Monique has payed considerable attention to how *ejidatarios* themselves attempt to (re)structure their lives and relationships with others and the state. In this way, they create space for their own projects which sometimes align with the state, and sometimes not. This does not mean that the state has no influence on local forms of organizing.

I take her approach to my study of an irrigation scheme in Zimbabwe. I likewise moved away from a strictly structuralist political economy framework (Hebinck, 2018) to explore the relationships between and amongst irrigators in Nyamaropa irrigation scheme. The analysis is set against the background of the debate in agrarian studies about how to interpret the social relationships between and amongst social actors (be they peasants, irrigators, or *ejidatarios,* traders, extensionists), markets, technology and the state (Hebinck and Van der Ploeg, 1997). Many scholars (Bernstein, 2010; Cousins, 2011; 2013; Scoones *et al.*, 2012) argue for a *class analytic* perspective. Other types of scholars like Dorward (2009) employ the prism of livelihood categories ('hanging in', 'stepping out') to capture transformation. Both models have in common to not take sufficient account of the richness of everyday life, the organising practices of social actors involved, their ways of solving of everyday problems, their organisation of their productive and consumptive resources, and thus of the ways how they shape social change and agrarian transformation. What is ignored is what I frame here as '*navigating everyday life*' which displays the local, organising practices of irrigators. Scholars like Vigh (2008), Cleaver (2002; 2012) and Cleaver and De Koning (2015) coined the term *navigating* to express how social actors patch social arrangements together '*from cultural resources available to them in response to changing conditions*'. How social actors navigate their social and economic environments and relate to the state through various kinds of practices that revolve around strategic distancing from markets and the state gets lost in political economy analyses (Hebinck and Van der Ploeg, 1997). I contend that these practices and relationships in turn reorder and reassemble existing agrarian structurers through engaging in *symbiotic* relations.

I argue, with Monique, that it is theoretically significant to unpack practices of navigation, which entails exploring how they deal with state policies and markets to structure their livelihoods and social relationships. The role of the state cannot be ignored as it plays a significant role in irrigation; if only through its engineers designing irrigation schemes and laying out how irrigators should cooperate to make use of irrigation facilities, maintenance of the scheme and so on (Zawe, 2006). These designs entail forms of cooperation between irrigators (e.g. sharing water, waiting for water turns) and disciplining by irrigation authorities (e.g. extensionists, engineers), but these relationships and the evolving constituting agrarian structure do not fully structure relationships. And, these relationships do not necessarily hinge on class and class relations *per se*. We need to be cognisant though of social actors being differentiated in terms of power, agency, and assets. But there are ways by which they manage to navigate these social and economic inequalities that characterise social life in irrigation schemes. There are those scholars, like myself, that alternatively suggest that social relationships are shaped by relations of *reciprocity* (Polanyi, 1957, 1992) and can be labelled and interpreted as *symbiotic* (Hill, 1963; 1969; Wegerif, 2017; Wegerif and Hebinck, 2016). These are socially significant and very often of a non-commoditised nature, allowing for exchange relationships that are not structured by the market and providing space for the enhancement of livelihoods that not are fully commoditised (Long, 1984).

One empirical context to capture the navigation practices is Nyamaropa Irrigation Scheme in north-east Zimbabwe. The history of the scheme provides context to picturing the changing conditions which are the product of the dynamics that (re) shape the political economy of Zimbabwe (e.g. colonial and post-colonial land policies, irrigation development and design, post-independent economic meltdown following the fast track land reform, the dollarization, record inflation, dysfunctional markets, recent state-orchestrated initiatives such as 'Command Agriculture') but – and that is the core of my argument – also what takes place at the level of the relationships amongst and between irrigators and traders, and how these shape the fortunes of irrigators.

I draw for my analysis on the few works published on Nyamaropa (Karadzandima, 2008; Magadlela, 2000; Magadlela and Hebinck, 1995; Reynolds, 1969; Satuku, 2017). These provide rich material to pursue an analysis of the evolution of agrarian relationships. During August 2017 and 2018, I revisited together with Cosmas Satuku and Sheila Chikulo some of the case studies of irrigators I did between 1995 and 2000 together with Duminsani Magadlela. Over the years, I managed to interact with about 70 irrigators across the scheme as well as with Agricultural Technical and Extension Services (Agritex) officers and traders, and marketing board representatives. I participated in extension field days, Irrigation Management Committee meetings and community gatherings and occasionally funerals.

A bit of history

The scheme was designed in the early 1950s as a gravity-fed scheme to produce food for consumption and exchange. Politically it provided space for black people who were forcefully removed from areas that were designated for white settlement elsewhere in what was then Rhodesia. The scheme became operational from 1961 onwards. From the beginning, the scheme was cultivated by 'newcomers' and by 'locals'. Reynolds (1969) made this distinction to differentiate between those that came from elsewhere (hence 'newcomers') and those that belonged to the original inhabitants of Nyamaropa and were governed by the headman Nyamaropa (hence 'locals). The majority of the 'locals' rejected the idea of irrigating, and only a few of them joined. It did not fit their lifestyles as they did not imagine themselves as the full-time farmers that irrigation required. They also resented paying irrigation fees for the land they considered theirs and accused 'newcomers' of taking over their land, and of being puppets to the colonial authorities who were confining people to poor and arid areas. The 'locals' were given the choice of joining the scheme as irrigators or moving off the land onto the surrounding hills. Most of them settled on the slopes of the hills. They preferred to structure their livelihoods around rain-fed agriculture and to combine this with working for the 'newcomers in exchange for some cash and a temporary use of a strip of irrigated land (Magadlela and Hebinck, 1995). The 'newcomers' and others that joined the scheme were registered as owners and were allocated on average 2 plots per family (0.8 ha in total); some were allocated 4 plots (1.6 ha.). With the allocation of a plot came water rights. The plots can be inherited by the sons, and if so needed be subdivided among the sons, and sometimes also daughters.

'Newcomers' and 'locals' were enmeshed in conflict for more than 30 years over formal access to the plots in the scheme (Magadlela, 2000; Magadlela and Hebinck, 1995). The conflict simmered despite that many 'locals' over time became involved in irrigation especially in winter and inter-marriages. The conflict was resolved in the end in an extension of the scheme in 2001 which created space for the 'locals' to also formally register as irrigators.

Ordering of the scheme: land-people relations

The schemes agrarian structure, so to speak, falls into three social categories differentiated by the type of relationships with others. These categories or vignettes should not be treated rigidly but as dynamic and with fluid boundaries. There are those that cultivate crops to generate a *profit* to *reinvest* in agriculture. Two other categories of analytical and numerical importance are irrigators sharing land, instruments, and labour rather symbiotically. These hinge on reciprocity, kinship, and religious affiliation. Others irrigate on their own accord and defend their *autonomy* by relying on their own resources, producing for own consumption and for a market but one that they prefer themselves. Common to all these types of arrangements is that they involve connections beyond the scheme extending into relations and exchanges with the dry-land farming community.

Farming for profit

The '*farming for profit*' arrangements fits with the *entrepreneurial* farmers in the scheme that invest their capital in cultivating land rented from fellow irrigators and dryland farmers. They engage in localised forms of contract farming. The '*for profit*' irrigators collaborate with government in programmes like 'command agriculture' (see Mazwi *et al.*, 2019). Some are full-time irrigators; others combine irrigation with off-farm jobs like teaching or trading, or they have access to remittances from family members. They manage to realise substantial profits that they reinvest in more resources for consumption and production and in other businesses.

Lovemore Nyabasa is such an entrepreneurial farmer. He combines teaching with farming. His two elder sons who are at university help him in the summer cultivating the land. Lovemore does not live close to the scheme but visits weekly to monitor his fields. He has large plots in the dry land only but also cultivates irrigated plots belonging to his relatives. He has eight oxen, six cows and three calves. The cattle stay in Mozambique where the 'pastures are greener'. He uses his oxen for ploughing his own fields and those of others in exchange for cash. The maize he produces – 6.5 tonnes of maize in 2018 – was under the Command Agriculture arrangement from which he received seed and fertiliser on credit. He sold 5,250 tonnes to the Grain Marketing Board (GMB) to repay the inputs. The remainder he sold on the local market.

Symbiotic arrangements

Typical for 'symbiotic' arrangements are not what commentators would expect: they produce for the market, and some of them quite a lot. Key to symbiotic farming and farmers is the sharing of resources (land, labour, seeds) to secure those for themselves as well as to prevent others to fall into poverty. They are not necessarily kin-related, and in many cases well-connected by religious affiliation. These symbiotic arrangements hinge as much as possible on non-commoditised relationships. Magadlela (2000) provides some detailed descriptions and analyses of what he calls a 'confederation of households'.

Steddy Nyagota cultivates a 0.8 ha irrigated plot next to a dry land plot. He works these plots with his wife, three sons and one daughter and a grandson. He owns two oxen, two cows and one calf. The family cultivates maize in the scheme and they rent dry land in the summer seasons in exchange for inputs or ploughing a piece of land. Those people he rents from do not have cattle or enough family labour, so they depend on others for ploughing and labour for planting. Johane Makunura can also be categorised as a symbiotic farmer. Johane and his two wives work the fields and employ most of the time their own labour. Occasionally they exchange labour with their relatives. They do this in turns. He uses his two oxen for ploughing his and others' fields. He has a 0.6 ha. plot which he inherited from his father and a 0.8 ha. plot in another part of the scheme. Johane also rents 0.4 ha from Masaya, which is important for the latter who has no other form of income besides farming. The rental fees Johane receives pays for water charges and production inputs of his remaining 0.4 ha. plot.

Individual

Jane Mautsa is a widow and cultivates 1.2 ha of irrigated land. She also rents another 0.8 ha. With the recent death of her husband, she lives a life of a fulltime farmer. Jane has 2 oxen for ploughing and 1 cow and grows a variety of crops. She relies on her own labour; occasionally she employs someone to help her in the fields and herding of cattle. Jane sells her produce to the GMB and to informal traders in the market centre. She does not participate in Command Agriculture which she does not like. She likes to be independent, make her own money and decisions. The money she makes from growing crops pays for her daughter's school fees, agricultural inputs and the fees for the plots and water.

Conclusion

This short article zooms in the agrarian structure in an irrigation scheme that is patterned on individual-based forms of accumulation in which the state and other agencies play a key role (e.g. 'Command Agriculture'). This perhaps being the intention of the state irrigation policies, new forms of irrigation community building and relationships emerged over the years hinging on reciprocity, kinship, and religion, whereby seed, land, water, and labour are shared reciprocally (but not always equally). I also identified irrigators that like to be independent or autonomous as much as possible. They rent land and hire labour but distance themselves where possible from state orchestrated arrangements to protect their autonomy and engage in exchange relations of their own choice. These three categories not only manifest different kinds of organising practices to manage their irrigation plots but also different patterns of social relationships and knowledge.

In my understanding of this kind of dynamics of social change and transformation, this 'structure' does not represent what Cousins (2013) refers to as an 'untransformed agrarian class structure', or what Bernstein (2015) sees as 'classes of labour residing in the rural areas'. Nor can we simply speak of 'stepping out' and or 'hanging in' as Dorward (2009) analyses differentiation processes. In contrast, I recorded active involvement in agricultural activities albeit in some form of sub-contracting relationships but also land, water and labour sharing relations across class and gender relationships. These relationships can only be properly grasped as embedded in and shaped by locally accepted and shared cultural repertoires. Such local forms of the organisation go beyond class *per se* and require a detailed ethnographic exploration of how social actors collaborate, deal with the state and create space for themselves by (perhaps) restructuring or creating a different kind of local and global political economy. Detailing and documenting the richness and complexities of everyday life is a theoretically significant aspect of such an analysis; one to which Monique Nuijten has devoted her academic career.

Saludo and *tatenda*.

Paul Hebinck / Emeritus Associate Professor, Sociology of Development and Change Group, Wageningen University, the Netherlands / Adjunct Professor, Agricultural Economics and Rural Extension, University of Fort Hare, South Africa / Senior Research Associate, Department of Environmental Science, Rhodes University, Grahamstown, South Africa / paul.hebinck@wur.nl

References

Bernstein, H., 2010. Class dynamics of agrarian change. Agrarian Change & Peasant Studies, Nova Scotia: Fernwood Publishing Company.

Bernstein, H., 2015. Commercial farming and agribusiness in South Africa since 1994. In: Cousin, B. and Walker, C. (eds.) Land divided. land restored. Land reform in South Africa for the 21st century. Aukland Park: Jacana, pp. 104-120.

Cleaver, F., 2002. Reinventing institutions: bricolage and the social embeddedness of natural resource management. The European Journal of Development Research 14(2): 11-30.

Cleaver, F., 2012. Development through bricolage: rethinking institutions for natural resource management. London: Routledge.

Cleaver, F. and De Koning, J., 2015. Furthering critical institutionalism. International Journal of the Commons 9: 1-18.

Cousins, B., 2011. What is a 'smallholder'? Class-analytic perspectives on small-scale farming and agrarian reform in South Africa. In: Hebinck, P. and Shackleton, C. (eds.) Reforming land and resource use in south africa: impact on livelihoods. London: Routledge, pp. 86-112.

Cousins, B., 2013. Smallholder irrigation schemes, agrarian reform and 'accumulation from above and from below' in South Africa. Journal of Agrarian Change 13(1): 116-139.

Dorward, A., 2009. Integrating contested aspirations, processes and policy: development as hanging in, stepping up and stepping out. Development Policy Review 27: 131-146.

Hebinck, P., 2018. De-/re-agrarianisation: global perspectives. Journal of Rural Studies 61: 227-235.

Hebinck, P. and Van der Ploeg, J.D., 1997. Dynamics of agricultural production. an analysis of micro-macro linkages. In: De Haan, H. and Long, N. (eds.) Images and realities of rural life. Wageningen perspectives on rural transformations. Assen: Royal van Gorcum, pp. 202-226.

Hill, P., 1963. Markets in Africa. Journal of Modern African Studies 1(4): 441-453

Hill, P., 1969. Hidden trade in Hausaland. Man 4(3): 392-409.

Karadzandima, L., 2008. 'No proper seed no bumper yield' smallholder seed sractices: a case study of Nyamaropa irrigation scheme, Manicaland, Zimbabwe. Wageningen: Wageningen University.

Long, N. (ed.), 1984. Family and work in rural societies: perspectives on non-wage labour. London: Tavistock.

Long, N., 2001. Development sociology: actor perspectives. London: Routledge.

Magadlela, D., 2000. Irrigating lives. Development intervention and dynamics of social relationships in an irrigation project in Zimbabwe. Sociology of Development and Change. Wageningen University, Wageningen, PhD thesis.

Magadlela, D. and Hebinck, P., 1995. Dry fields and spirits in trees – a social analysis of development intervention in Nyamaropa Communal Area, Zimbabwe. Zambezia 22(1): 44-63.

Mazwi, F., Chemura, A., Mudimu, G.T. and Chambati, W., 2019. Political economy of command agriculture in Zimbabwe: a state-led contract farming model. Agrarian South: Journal of Political Economy 8(1-2): 232-257.

Nuijten, M., 1998. Organization, transnationalism, and the culture of the state in a Mexican Ejido. Wageningen: Wageningen University.

Nuijten, M., 2003. Power, community and the state: the political anthropology of organisation in Mexico. London: Pluto.

Polanyi, K., 1957. The great transformation: the political and economic origins of our time. Boston: Beacon Press.

Polanyi, K., 1992. The economy as instituted process. In: Granovetter, M. and Swedberg, R. (eds) The sociology of economic life. Boulder: Westview Press, pp. 29-51.

Reynolds, N., 1969. A socio-economic study of an African development scheme. Cape Town: University of Cape Town.

Satuku, C.T., 2017. A socio-economic history of Nyamaropa communal area with particular reference to agriculture, 1960-1978. Harare: University of Zimbabwe.

Scoones, I., Marongwe, N., Mavedzenge, B., Murimbarimba, F., Mahenehene, J. and Sukume, C., 2012. Livelihoods after land reform in Zimbabwe: understanding processes of rural differentiation. Journal of Agrarian Change 12(4): 503-527.

Vigh, H., 2008. Crisis and chronicity: anthropological perspectives on continuous conflict and decline. Ethnos 73(1): 5-24.

Wegerif, M., 2017. Feeding Dar es Salaam: a symbiotic food system perspective. PhD thesis. Rural Development Sociology. Wageningen: Wageningen University.

Wegerif, M. and Hebinck, P., 2016. The symbiotic food system: an 'alternative' agri-food system already working at scale. Agriculture 6(3): 40.

Zawe, C., 2006. Reforms in turbulent times. a study on the theory and practice of three irrigation management policy reform models in Mashonaland, Zimbabwe. Wageningen: Wageningen University.

5. POWER, COMMUNITY AND THE STATE: AN ENDURING CLASSIC OF ANTHROPOLOGY AND DEVELOPMENT

David Lewis

With this short note of appreciation I wish to pay tribute to Monique Nuijten's book *Power, Community and the State: The Political Anthropology of Organisation in Mexico*, in recognition both of its contribution to the field of anthropology and development and the multiple ways the book has informed, and enlarged my own ideas. I have always kept Monique's monograph close. It touches on so many of the themes I have explored in my own work as part development studies researcher, part anthropologist. Some of its influences will be obvious, others perhaps less so.

I have come to know Monique and her work mainly through the EIDOS discussion meetings (mentioned elsewhere in this volume by Oscar Salemink). Over the years these informal meetings have always felt worthwhile – productive and fun. They made a welcome difference to one's academic life by bringing together, usually without any particular agenda, expectations or pressure, a small group of people to discuss ideas. They almost always led somewhere interesting, and even more importantly, succeeded in humanising intellectual life in ways that the day to day pressures of one's own university experiences can all too often obscure. A good part of that was due to Monique's warm personal presence and constant readiness to share and exchange ideas.

But back to the monograph. In *Power Community and the State* Monique's starting point was to build on Norman Long's actor-oriented sociology approach to the sociological study of development. Picking up the challenge arising from Eric Wolf's (1990) observation that anthropologists had by this time largely 'relinquished the study of organisation', Monique sets about remedying this gap by extending this actor oriented approach to studying organisations as they actually operate on the ground. This commitment to understanding the realities of ordinary lives is of course completely in keeping with Monique's personality and her way of being in her wider life – the personal and the professional are always fully aligned. She does this because she wants us to see beyond the tendencies of much of the development literature (as well as the imperatives policy makers) to see organisations simply as 'instruments of change' and the assumption that organisations can somehow be re-engineered to bring about desired changes in the lives of poor people. At the heart of her analysis is an emphasis on the structural power that drives organising processes and this has always been something that I have tried to put at the heart of my own work on NGOs and civil society in Bangladesh. Monique's careful attention to theory and fine-grained ethnography is one of the best illustrations of how to do this.

The book engages with so many themes, from big picture concepts of governance and 'corruption' to the unpacking of the ethnographic ground level details of local practices of organising. For example, Monique describes how violence, corrupt practices and patron-client relationships help to constitute state power as a 'shadowy realm' that lies beneath, a perspective that has only grown more persuasive and become reflected in a growing body of other work that takes a post structural approach to understanding the state (such as Jackman, 2019; Lund, 2006). The insight that anti-corruption discourses do not undermine the Mexican political regime but instead help to constitute it, is of course a lesson that still needs to be learned by many development policy makers and NGO activists. The detailed ethnographic material that is contained in Monique's account describes how the power of gossip, speculation and rumours is constitutive of power at the interface between people and the state. This perspective, from the vantage point of our social media age almost two decades on, is of course another theme that is becoming even more relevant in today's politics.

Also important is the way Monique's monograph traces the ways that power is exercised through the *simplifications* of policy. She both draws upon and adds to James Scott's (1998) portrayal of the power of 'state simplifications' that serves to narrow the vision and make certain phenomena more 'legible' and therefore more controllable. This was a ground breaking idea two decades ago and has grown in significance since, as an important theme for anthropology and social science more widely. Organisational reform initiatives, she writes, 'by themselves have little chance of changing existing power relations and bringing more prosperity to the poor. This instrumental view of organisational reform leads to a vicious circle in which ill-functioning organisations are made the scapegoat for the bad socio-economic conditions of the poor, and against which the propagation of new organisations is used as a magic charm' (Nuijten, 2003: 190). Organisations are spaces, or 'empty shelters' (*ibid.*:190), as she puts it, where relations of power are all too often masked, and where modernist notions of accountability make very little sense.

Finally, and perhaps less obviously, Monique's analysis is important because it refuses the simplistic binary geographies of the Global North and the Global South, remarking that 'there is less difference in the nature of politics in different places than is normally assumed' (p.205). Problems of governance or unaccountable leadership are commonly attributed to the dysfunctional political systems of 'developing' countries, but as she argues, they should not be. This is an important insight from the book that has perhaps become truer with time. For example, journalist and activist George Monbiot (2020) recently remarked that London is one of the most corrupt cities in the world because 'kleptocrats in other countries are merely clients of the bigger thieves in London'. In my own work, I have examined the ways these unhelpful binaries (North/South, developed/developing, etc.) tend to obscure the ways power operates globally, and the productive potential that can sometimes be generated for those who attempt to question, challenge and transcend them (Lewis, 2011).

Monique's work continues to inspire because it has stayed relevant and its ideas set a standard towards which many of us continue to aspire. Monique offers an example of an approach to academic work that is both rigorous and people-centred, personal and political, and theoretically informed while always remaining open to detail, nuance and above all to a unique form of human warmth and openness.

David Lewis / Professor of Social Policy and Development, London School of Economics and Political Science, London, United Kingdom / d.lewis@lse.ac.uk

References

Jackman, D., 2019. The decline of gangsters and politicization of violence in urban Bangladesh. Development and Change 50(5): 1214-1238.

Lewis, D. and Hossain, A., 2019. Local political consolidation in Bangladesh: power, informality and patronage. Development and Change in press. https://doi.org/10.1111/dech.12534

Lewis, D., 2011. Tidy concepts, messy lives: defining tensions in the domestic and overseas careers of UK non-governmental professionals. Chapter 9. In: Mosse, D. (ed.) Adventures in Aidland: the anthropology of professionals in international development. London: Berghahn Books.

Lund, C., 2006. Twilight institutions: public authority and local politics in Africa. Development and Change 37(4): 685-705.

Monbiot, G., 2020. If you think the UK isn't corrupt, you haven't looked hard enough. The Guardian, September 10. Available at: https://www.theguardian.com/commentisfree/2020/sep/10/uk-corrupt-nation-earth-brexit-money-laundering

Nuijten, M., 2003. Power Community and the State: The Political Anthropology of Organisation in Mexico. London: Pluto Press.

Scott, J.C., 1998. Seeing like a state: how certain schemes to improve the human condition have failed. New Haven: Yale University Press.

6. CITIZENSHIP AND POLITICAL AGENCY IN LATIN AMERICA

Sian Lazar

Monique Nuijten's work in Mexico and Brazil speaks clearly to contexts that I know well in Bolivia and Argentina; but also across much of Latin America and more widely. Her book *Power, Community and the State* can be read in at least two ways: as an ethnography of development and as an ethnography of the state. Both are very big topics, so I would like to focus on the second, and pull out some of the insights that have been most striking to me. Through effective, sensitive and meticulous ethnography of Mexican *ejidatarios* (peasants, on formerly collective landholdings), she shows how people have a deeply contradictory relationship to the state. On the one hand, the state is provider of resources, protection, jobs (perhaps); while on the other, it is a deeply corrupt enemy; it is both protector and oppressor. In a beautiful turn of phrase, she says that bureaucracy functions to create a 'never ending cycle of high expectations followed by disillusion and ironic laughter' (Nuijten, 2003: 197). All this produces the fantasy of a good state alongside the cynicism that comes from knowing that in reality it will not come to pass. And yet, they/we keep going. This is the bureaucracy as 'hope-generating machine'. For Monique, the state is not the effective disciplining machine implied by some of the governmentality literature, and yet as all its disparate entities join together they create and reproduce hope of something better. It is a form of enchantment that bolsters a strong culture of the state, even when the state apparatus itself is not especially effective or even benign.

One of the most important ways that this happens is through personalised politics, especially associated with patronage and clientelism. This theme crosses both her book and her citizenship article (Nuijten, 2013). In her book, she shows how politics and bureaucracy in Mexico are utterly personalised, which is partly how corruption comes to make sense. It's not a dysfunction in a system, nor the result of individual 'bad apples', but an effect of the state's humanity and its contradictions. So, she argues, scholars have not paid sufficient attention to the psychic aspects of corruption: the pleasure and enjoyment that comes from doing someone a favour; the affective charge of connections; the need to know people in order to get something done and what that means for agency. In such a system, it should be possible to get that problem resolved, to disentangle the legal knots that have left your community without land it thought it controlled. You just need to find the right connection to mobilise, the right person to know. After all, that other problem got resolved that other time, so why should it not work again? This personalised nature was especially evident under the political system in Mexico at the time, as the PRI monopoly of power was coming to an end, but I think that it's a brilliant description of how politics works in the region and not peculiar to Mexico. (I suspect that it's the case also for Europe, it's just that we are much less willing to see it.)

Patronage appears as key also in the article Monique wrote for the special issue of *Critique of Anthropology* that we co-edited. By now, she had moved to work in Brazil, and the article is about a slum-upgrading project in Recife – '*The perversity of the 'Citizenship Game': Slum-upgrading in the urban periphery of Recife, Brazil*' (2013). Here, she identifies the contradiction between the citizenship language of the local government officials from the PT (Lula's party), and the ways that people themselves conceptualise the scheme. They speak two different 'languages of the political' in Monique's words: the officials speak a language of (neoliberal) citizenship and participatory development, where individuals make a better life in conjunction with government but without intermediation from brokers, while the slum dwellers speak a language of patronage, where new houses are a gift from government to which the appropriate response is gratitude, and the knowledge that you reciprocate by voting for the right person. Monique tells us that 'the language of citizenship tends to be framed around 'rights' and 'duties' while the language of patronage is organised around 'gifts', 'taking care', and 'exchange of favours' (2013: 21). In fact, plenty of project officials use the second language when needed, especially as the elections approach, but their driving discourse in their policy materials and the meetings they hold emphasises how the people of the slums should be good citizens and fulfil their duties. I think that where she and I would differ conceptually is that for me both are practices of citizenship, even if only the former is explicitly so.

Meanwhile, the people who receive new houses have a complicated response to the project: they think the houses are more hygienic, and the district more 'cidade' or 'city' (as opposed to slum or favela). Also, having an address suitable for an electricity service (and bill) is considered a good thing because it enables people to access formal consumer credit. On the other hand, the electricity is costly in comparison to their previous informal (illegal) arrangements, and the houses are too small and insecure. Residents complained about the grid layout of the new district, which, they said made it 'easier for gangs and police patrols to drive through, shoot at targets, and get out again' (2013: 18); and they quickly installed bars on their windows and doors when they could. Monique concludes that the 'citizenship game' played by the state in these Recife slum resettlement projects is perverse in two ways. First, because it requires slum dwellers to become something other: self-disciplining citizens who will participate in a project on its own terms but without having any influence over it; and second because despite the rhetoric the project in fact stimulated patronage. So, the citizenship aspect of the resettlement project failed in its two main aims. The slum dwellers actually did not comply with the kind of citizenship expected of them, but ultimately, neither did the project. As in Mexico, people respond to government and power in complex and contradictory ways.

The gritty urban environment of the slums of Recife contrasts with the rural Mexico of *Power, Community and the State*, but Monique's clarity of writing, ethnographic sensibility and subtle empiricism infuses both analyses. They show the value of an anthropological approach that pays close attention to what people say, how they feel, and what they do as they adapt their lives to state projects, beyond boring dichotomies of accommodation and resistance. In Monique's hands, that then brings out the affective, phantasmatic and enchanting aspects of the state in these specific contexts, which then prompt the rest of us to think about these aspects in the places where we work. Both pieces are models for how we should carry out political anthropology.

I have engaged with Monique as a reader and a colleague. We collaborated in the organisation of a workshop (in 2010) and the resulting journal special issue (published in 2013) and what I mostly remember about that is how easy it all was. I can't even remember how it came about, just that somehow we ended up co-funding this workshop and that Monique was incredibly easy-going about the whole process. As we pulled together the workshop and the publication, she was supportive and calm, but also intellectually challenging and insightful. I'm proud of the special issue that we co-edited, which brought together questions of citizenship and political agency. I don't think that she and I completely agreed about how to use citizenship conceptually, but we had very similar views about the nature of the relationship between people and state. It was enormously enriching – and fun – discussing all these kinds of questions with Monique during the workshop, and it has been a brilliant experience to go back to these publications as I write this piece. Thank you.

Sian Lazar / Reader in Social Anthropology, Department of Social Anthropology, University of Cambridge, United Kingdom / sl360@cam.ac.uk

References

Nuijten, M., 2003. Power, community and the state: the political anthropology of organisation in Mexico. London: Pluto.

Nuijten, M., 2013. The perversity of the 'Citizenship Game': Slum-upgrading in the urban periphery of Recife, Brazil. Critique of Anthropology 33(1): 8-25.

7. NEGLECT AND THE IMAGINATIONS OF CARE; SOME BRIEF THOUGHTS INSPIRED BY MONIQUE NUIJTEN'S WORK

Julia Eckert

There is something I want to discuss with Monique, an observation of hers that I find very perceptive, and that to me still bears possibilities for further interpretation. Monique frequently pointed to her observation of people feeling *neglected* by the state. In her text on 'Ritual and Rule in the periphery', that Monique Nuijten wrote with David Lorenzo (Nuijten and Lorenzo, 2009), for example, they begin with the saying of people that the state was to them like a 'mother that rejects her children'. It is an absent state, one that is present rarely, and if, then only through violence. These experiences of neglect and abandonment by the state of the people she worked with is something that appears in several of Monique's texts, and it has stuck with me, because it appears as a very poignant observation, one that I would love to discuss with her further.

The neglectful state abandoning those subjected to it that we encounter in Monique's texts is one not unfamiliar to many of my interlocutors, who, like the Peruvians Monique met, express the same experiences of neglect. In fact, 'neglect' appears as one of the most ubiquitous experiences people articulate with regard to their relation to the state administrations that govern them, but one that has rarely been explored in depth. It speaks of an expectation that has been disappointed, a promise not kept. This specific affect of 'feeling neglected by the state' appears to emerge in vastly different situations and thus speaks of diverging wants; it relates to various disappointments and absences, indeed: refusals on the part of state agencies: Lack of material provision, of physical security (Goldstein, 2012), lack of participatory possibilities, or of the sheer presence and responsiveness of state agencies.

Experiences of neglect seem to be intensifying in times when governance has visibly moved to multiple powers, and when it is often particularly marginalised populations that seek protection by the state from the unaccountable sway of multinational corporations and other forces and movements that are experienced as threatening and uncontrollable. Seeking refuge in a centre of power often harks back to an idealised past, or to mythical forms of pastoral authority, the Peruvian mother mentioned by Monique, *Vater Staat* or *devlet baba*.

We might read such expressions of feelings of neglect to index a desire for care beyond the allusions to parental authority in such expressions.

The biopolitical practices of state care have been analysed for the violence that they entail (e.g. Gupta, 2009). As anthropologists we have often tried to understand how people try to navigate that violence, how they carve out their spaces of autonomy, and how they evade such violence by developing alternative modalities of care. Some have suggested that such alternatives would rely on and necessitate entirely different forms of organising our polities, would need to abandon 'the state' as an idea of a promise, such as the people James Scott (2009) has described as designing their whole social structure for the purpose of evading the state(s) that seek to control them. Echoing Clastres' (1974) analysis of the politics of Amazonian people as directed against the emergence and consolidation of any central authority, such analyses suggest that resistance and refusal (Prasse-Freeman, 2020) are the two modes in which true transformation appears as possible, because they overcome the hegemony, the cultural power of the state-idea that keeps a hold on our desires and goals. In such a perspective, 'neglect' points to a mode of relating to state agencies that falls between the rock of hegemony and the hard place of resistance; expressions of feeling neglected signal the opposite of refusals and resistances. They want more. They articulate a disappointed expectation, a call to 'a mother that rejects her children', and point to a violated obligation. One could see this merely as the sign of an utter *étatisation* of minds, in the sense that the desires and expectations underlying feelings of neglect reproduce statist understandings of people's relation to authority, statist definitions of needs and allocations of responsibility. The hegemonic image of the state might make it impossible to imagine a different state, and experiences of neglect are then merely the expression of the disappointment about the divergence between what was promised and what is actually given. One might thus say that what we see in 'neglect' is the inability to think beyond the particular notion of the public good and the biopolitical affordances that states make towards their citizens at a certain historical moment.

There is something to that. People do claim what is promised, trying to hold the state by its word, and condemning it for its breaches of promises (Eckert, 2006); this is evident in the turn to the language of rights and the juridification of protest, since they mean entering a language that states cannot entirely ignore. However, this might not be all to it. As I have argued elsewhere, the interpretations (of law) that occur in practices of citizenship, translate needs and desires into claims on the state, and thereby articulate political norms about the way the state should act, not necessarily confined to the statist norms of care, but transposing values and experiences from elsewhere into their demands. Furthermore, Monique's insights are extremely helpful to understand that also in situations in which no promise, let alone any 'social contract', had ever been agreed upon or even offered (Bear and Mathur, 2015), people desire care from the state and responsiveness to their needs. Their expressions of neglect are like a demand, a claim, and in this claim they generate an image of what should be, a normative suggestion and goal.

What is missed out on if such expressions of neglect are discounted as mere effects of the *étatisation* of minds, are the political theories inherent in them, and that within such political theories that possibly exceeds the hegemonical norm.

Maybe our inability (or refusal) to fully theorise these disappointments with the state and is due to even our disciplines and notions of critique being strangely stuck in a romantic notion of freedom (Englund, 2006) in relation to politics. Both Harri Englund (2006) and James Ferguson (2013) have pointed at 'dependence' as a political mode and value. Its positive valorisation in relation to the state is well-nigh inconceivable for good reasons, given the way dependence always goes along with the power of control, surveillance and sanction on the side of the 'patron'. However, as both Englund and Ferguson suggest, dependence is also a claim towards an obligation of care, which complicates matters of freedom for all conjoined within relationships of dependence and care. Hence, if we think dependence and care together, and take feelings of neglect to express ideas of a caring polity, we could approach the pastoral state from the implicit political theories of those cared for and cared about. This might bring into focus the imaginations of what a good state would be that run through such feelings of neglect, and that exceed the models of care provided by the state, possibly drawing on norms and values experienced in other relations and contexts. I would venture the thesis that in most expressions of neglect we do not see (only) the disappointment over a broken contract, but actually the suggestion for alternative ways of government. 'Neglect' bears within it references to alternative imaginations and other norms; it signals not bare dependence but rather demands that carry multivalent aspirations. 'Responsiveness' more aptly describes what seems to be at issue, and responsiveness entails both provisions of care for the needs expressed and participatory possibilities, if only to express what needs matter. If we thus take these expressions of the experiences of neglect seriously, rather than seeing them as the state effect of a hegemonic norm of how to imagine social order, they might signal politics 'otherwise', the otherwise not being alternatives *to* the state but alternative states (see also Cooper, 2019).

Julia Eckert / Professor of Social Anthropology, University of Bern, Switzerland / julia.eckert@anthro.unibe.ch

References

Bear, L. and Mathur, N., 2015. Remaking the public good: a new anthropology of bureaucracy. Cambridge Journal of Anthropology 33(1): 18-34. https://doi.org/10.3167/ca.2015.330103

Clastres, P., 1974. La société contre l'état. Recherches d'anthropologie politique, Paris: Minuit.

Cooper, D., 2019. Feeling like a state; desire, denial, and the recasting of authority. Duke University Press.

Eckert, J., 2006. From subjects to citizens: legalisation from below and the homogenisation of the legal sphere. The Journal of Legal Pluralism and Unofficial Law 38(53-54): 45-75.

Englund, H., 2006. Prisoners of freedom; human rights and the African poor. University of California Press.

Ferguson, J., 2013. Declarations of dependence: labour, personhood, and welfare in southern Africa. Journal of the Royal Anthropological Institute 19(2): 223-242. https://doi.org/10.1111/1467-9655.12023

Goldstein, D., 2012. Outlawed: between security and rights in a Bolivian City. Duke University Press.

Gupta, A., 2012: Red tape; bureaucracy, structural violence, and poverty in India. Duke University Press.

Nuijten, M. and Lorenzo, D., 2009. Ritual and rule in the periphery: state violence and local governance in a Peruvian comunidad In: Von Benda-Beckmann, F., Von Benda-Beckmann, K. and Eckert, J. (eds) Rules of law and laws of ruling. On the governance of law. Law, Justice and Power. Aldershot: Ashgate Publishing Limited.

Prasse-Freeman, E., in press. Resistance/refusal: politics of manoeuvre under diffuse regimes of governmentality. Anthropological Theory. https://doi.org/10.1177/1463499620940218

Scott, J., 2009: The art of not being governed: an anarchist history of Upland Southeast Asia. Yale Agrarian Studies Series.

8. ESPERANZA Y POLÍTICA

José Luis Escalona Victoria

Recuerdo que cuando estabas en México, Monique, se discutía sobre las consecuencias de un giro aparentemente radical en la política nacional, una supuesta transformación profunda del 'estado'. Tu trabajo, a la luz de esas discusiones, es todo un reto al pensamiento, uno de enorme valor heurístico. Permíteme explicarme.

En la segunda mitad del siglo XX muchos analistas se interesaron en las particularidades del Estado mexicano, esas que parecían distanciarlo de otros en América Latina. México mantuvo una fachada institucional estable, con un régimen de reglas democráticas que reciclaban a la élite gobernante cada tres y seis años (aunque ocurría, en general, dentro de un solo partido). Tenía también un programa supuestamente expresión de la Revolución Mexicana de inicios del siglo, con ideas de democracia, justicia social y nacionalismo. No obstante, esa autoimagen no daba cuenta de cómo operaba este estado, lo que desató diversas discusiones que entretuvieron a los académicos por varias décadas.

Por un lado, el estado de la revolución mexicana, si era tal cosa, era analizado por su ideología (por ejemplo, Córdova, 1973); no obstante, también era visto como un aparato carente de ideología (Paz, 1978). Para algunos analistas el estado era un aparato de violencia, más abiertamente en actos de represión contra, por ejemplo, los movimientos de trabajadores, de estudiantes, y la movilización guerrillera de los años sesenta y setenta; no obstante, también era un aparato de alianzas y colaboración de masas, lo que le permitió distanciarse de la ola de dictaduras que vivió América Latina (Woldenberg, 1988). La centralidad de la violencia o el consenso, de la ideología o de la práctica, dependían del punto de observación, ya fueran los casos de violencia o las amplias áreas de interacción entre gobierno y pueblo (con todas sus diversidades) en educación y salud, o de atención a 'problemas' específicos como el reparto agrario y el indigenismo (programas especiales para la integración de la población india), o la gestión de bienes nacionales (principalmente la industria petrolera – ver Lomnitz, 2001).

Por otro lado, se hablaba de una debilidad de las instituciones formales, lo que llevó a los estudiosos a explorar áreas de la política más allá de las normas o los programas. Desde un punto de vista legalista radical, la vida cotidiana parecía depender de prácticas profundamente 'ilegales' o 'corruptas', de relaciones y acciones que dependían de conexiones personales, hombres fuertes o fuerzas en la sombra, con leyes e instituciones sólo como fachada. Se hablaba así de caciquismo, caudillismo, corporativismo, presidencialismo y clientelismo (por ejemplo, de la Peña, 1986; Knight y Pansterns, 2006). No obstante, los diagnósticos eran muy diversos: por un lado, se hablaba de actores poderosos que usan las instituciones para imponer sus intereses; por otro, se negaba la eficacia de esas instituciones, o la existencia misma de un poder suficientemente organizado y centralizado. Analistas en la política, la academia y los medios de comunicación compartían ciertas ideas y cierto lenguaje sobre política y gobierno, disputando enfoques, énfasis e interpretaciones en la definición del estado.

Cuando llegaste al sur de Jalisco, decía, estábamos en medio de un giro aparente, pues muchos programas relacionados con el régimen de la revolución mexicana estaban siendo desmantelados, redefinidos o redirigidos. El giro era suave y su vórtice atrajo muchas confrontaciones políticas, incluso al mundialmente conocido Ejército Zapatista de Liberación Nacional. La discusión reanimó de hecho figuras retóricas, como las del pueblo pragmático, 'el México bravo', y la larga historia de política informal; también imágenes de un gobierno de personajes corruptos, violentos, o manipuladores, vinculados a poderosos empresarios y políticos. Muchos investigadores, algunos apasionadamente, se sumergieron en este debate.

En esas circunstancias, Monique, nos mostraste con tiento una interesante forma de revisar esos temas sin quedarnos atrapar por el politizado lenguaje mexicano ni por sus clichés sobre el estado. Por el contrario, nos presentaste una visión fresca, de alguien que mira y escucha de una manera más detallada y respetuosa a la política en su vida cotidiana, en sus múltiples idiomas y actuaciones, tomando nota aguda de cada cosa en su sitio, para analizarlas todas como prácticas y narrativas tal como surgen, como parte de flujos de acción y que son, por ello, cambiantes, situacionales, inciertas y contradictorias. Hasta donde he podido entender, tu libro, *Power, Community and the State, The Political Anthropology of Organisation in Mexico* (Nuijten, 2003), muestra una manera diferente de estudiar política y estado. Permíteme frasear aquí un poco de lo mucho que aprendí leyéndolo, o escuchándote en clases y conversaciones durante tu estancia en Michoacán.

Propones, primero que todo, partir de las *prácticas organizativas*. Pero antes que separar lo formal y lo informal, o lo legal y lo corrupto, en el sentido convencional, propones identificar pautas en el flujo de la acción. Parece una idea sencilla, pero esconde un acto de ruptura epistemológica mayor en la antropología política en México, pues implica dejar atrás ciertas ideas, poner atención en interacciones que son, en sí mismas, abiertas a diversos y contradictorios resultados, e identificar pautas.

Primero, como lo muestras en tu estudio en ese pueblo del sur de Jalisco (realizado entre interminables disputas por tierras, contrastantes narrativas de la 'revolución mexicana', disputas familiares, acusaciones de corrupción y teorías de conspiración) esos flujos de acción no son prácticas totalmente preconstruidas o determinadas por posicionamientos claros. Lo que ocurre son despliegues de negociación cargados de historias contradictorias y cambiantes, no sólo por parte de las personas que acuden a las oficinas gubernamentales sino también de los empleados de estas instancias burocráticas. Hay claro diferencias entre los que participan, que dan forma a campos de poder y que reproducen jerarquías, pero esas condiciones se reciclan en cada nueva fase o evento de negociación. Los campos de poder (la formación de jerarquías y diferenciales de poder) no son andamiajes fijos o preestablecidos sino resultado eventual de la historia de esas interacciones, relevantes sólo en las luchas concretas por recursos específicos.

Otro aspecto es su importante carga de teatralidad y discursividad. Las acciones derivan de y conforman negociaciones y cambios estratégicos. Por ello, parte de las pautas de las prácticas organizativas son las contradicciones, el caos, e incluso las mentiras y las incertidumbres (esa sensación kafkiana), expresadas de distintas formas tanto en la interacción misma como en los relatos posteriores de los participantes. El arte etnográfico, entonces, consiste en entender las actuaciones y las narrativas en distintos registros, incluyendo los chismes, las teorías de conspiración, el cinismo (sabiendo, por ejemplo, que pedir o dar un pago por un favor podría no resolver el asunto) y el movimiento estratégico (como con la búsqueda del conecte adecuado, el intermediario correcto con el centro todopoderoso del estado).

Otro aspecto destacable es la transmutación de ciertos objetos en fetiches, como los documentos y mapas oficiales (y quizás, extendiendo abusivamente el término, el mismo 'intermediario adecuado'), pues es en algún punto de los flujos de acción esos objetos o personas terminan siendo concreción de una idea de un poder estatal centralizado (una *idea de estado*). En contraste, las prácticas organizativas muestran que lo que hay es un conjunto de instancias descentradas, que no operan como extensiones de un orden central (personalizado o reglamentado) y que más bien siguen lógicas inciertas y contradictorias.

La política es entonces todo ese despliegue teatral y retórico, con movimientos estratégicos, cargados de cinismo, sorpresa e incertidumbre; no obstante, su devenir renueva el interés por estar involucrado en el juego. Una idea que sintetiza muy bien la perspectiva está en la noción de *máquina generadora de esperanza (hope-generating machine)*. El conjunto de prácticas organizativas permite entender de manera concreta cómo se enfrentan y se tratan de resolver problemas específicos, por ejemplo, un pleito agrario irresuelto por años pero que se podría resolver algún día. Las interacciones dirigidas a ese objeto, que involucran mapas, documentos, promesas, dinero, viajes a oficinas, reuniones con empleados, visitas personales de empleados, etc. son de algún modo el estado, que se renueva reactivando cada vez la esperanza. Se trata de una antropología que explora una forma distinta de entender la política, más allá, por ejemplo, de las ideas de violencia y terror (Taussig, 2000; Graeber, 2006) o de despolitización (Ferguson, 1994).

Finalmente, de este análisis surge una poderosa tesis, cuando se refiere a la *idea de estado*. Se trata de invocaciones a un poder centralizado que lo puede resolver todo, una figura retórica que reaparece en las charlas y las ceremonias, tanto entre los que buscan o se autopresentan como el 'conecte adecuado', como entre los que sugieren una conspiración detrás de todas las fallas o los fracasos, o en los críticos pertinaces del gobierno; tanto en las actuaciones y el habla, como en los objetos/fetiches. Lo interesante es que no sólo es pauta, por así decirlo, en las prácticas organizativas, sino también en la propia auto-representación de las instituciones gubernamentales, entre los empleados que se presentan como los que sí tienen el acceso directo con el poder (el gobernador o el presiente), e incluso en los medios de comunicación y en muchos análisis en ciencias sociales. Resulta inquietante pensar que incluso los análisis podrían haber estado atrapados en la fascinación por el estado, en su magia; es inquietante pensar que las discusiones académicas podrían ser, en cierto grado, productoras de la idea de estado.

Es una lástima que no exista una traducción de tu libro al español, aunque eso no ha sido un obstáculo para su difusión en el ámbito académico. Tomar en serio tus propuestas, Monique, aprovechar con más profundidad tu mirada antropológica, nos llevaría seguramente a repensar la literatura sobre política y estado en México. Tus propuestas, que aparecen ya recuperadas en diversos trabajos, son parte de una transformación en la investigación en áreas de antropología política, de la organización y del estado, así como en la práctica de la etnografía en general. Finalmente, los puntos críticos apuntados sobre la *idea de estado*, en esta era de creciente cinismo político, nos harán revisar con más cuidado nuestras premisas en el análisis y la discusión sobre política y estado en México.

No sé si he entendido todo; faltan charlas, Monique, como las de México, la que tuvimos brevemente en Manchester o las de Utrecht y Wageningen (cuando me hospedaste tan amablemente). Lo que sí se, es que he aprendido mucho y que puedo aprender más de ti. Por lo pronto, recibe estas palabras con mi amistad, admiración, respeto y agradecimiento.

Un abrazo.

José Luis Escalona Victoria / Full Professor, Centro de Investigaciones y Estudios Superiores en Antropología Social, Unidad Sureste, Chiapas, México / joseluisescalona@prodigy.net.mx

Bibliografía

Córdova, A., 1973. La ideología de la revolución Mexicana: la formación del nuevo régimen. México: Ediciones Era.

De la Peña, G., 1986, Poder local, poder regional: perspectivas socioantropológicas. In: Padua J. and Vanneph, A. (eds) Poder local, poder regional, México: El Colegio de México / Centre d'Etudes Mexicaines et Centramericaines, pp. 27-56.

Ferguson, J., 1994. Anti-politics machine: development, depoliticization, and bureaucratic power in Lesotho. University of Minnesota Press.

Graeber, D., 2006, Dead zones of the imagination. On violence, bureaucracy, and interpretive labor. The Malinowski Memorial Lecture.

Knight, A. and Pansters, W. (eds), 2006. Caciquismo in twentieth-century Mexico. Institute for the Study of the Americas.

Lomnitz, C., 2001. Elusive property: the personification of Mexican national sovereignity. In: Myers, F.R. (ed.) The empire of things. School of American Research Press, pp. 119-138.

Nuijten, M., 2003. Power, community and the state. Pluto Press.

Paz, O., 1978. El ogro filantrópico, México. Revista Vuelta 21.

Taussig, M., 2000. Un gigante en convulsiones. El mundo humano como sistema nervioso en emergencia. Barcelona, Gedisa.

Woldenberg, J., 1988. La lucha por la Hegemonía en México. México, Siglo XXI.

9. ANTHROPOLOGY OF THE STATE AND CONTEMPORARY STATE-BUILDING

Georg Frerks

Introduction

Being part of the same Department of Rural Development Sociology at Wageningen University during roughly two decades, Monique Nuijten worked largely on the anthropology of the state, particularly with reference to Mexico, while I worked on disaster and conflict with a particular interest in the Sri Lankan conflict, in a separate section called Disaster Studies. Though our ways crossed regularly, sitting in the same corridor, and we appreciated each other as good colleagues and friends, our areas of work proceeded along different paths. As a consequence, we knew each other's work in a broad and more general, but not very in-depth way.

The writing of a piece for Monique's *Liber Amicorum* was therefore a good opportunity for me to re-acquaint myself with her work, especially that in relation to the state. Reading through her book *Power, Community and the State: The Political Anthropology of Organisation in Mexico*, I was struck by a number of her insights, and it immediately occurred to me how relevant they were to current international debates on external intervention in state-building. In state-building an explicit attempt is made to design, set up or strengthen a state or its institutions by outside intervention. In my own work

I have been repeatedly investigating, teaching and writing on western military and civilian interventions in post-conflict societies: what they tried to achieve, and how they frequently were mal-designed or mal-executed, if not mal-intentioned.

In what follows, I take a few key notions from Monique's work and try to reflect on their relevance in this context. I shall attempt to do this by relating them to some problematic aspects of contemporary state-building practice, and then to link them to some pertinent academic and policy debates around these subjects by selected authors.

Some key insights from Monique Nuijten's work

In a short piece like this, it is not possible to do justice to the depth and breadth of Monique's work. At the same time, just taking a few interesting insights from it is also a risky affair. One easily takes insights out of context or applies them to a different or less fitting situation or case. I believe it is worthwhile taking that risk, as such an effort can provide fertile ground to explore new avenues, shape new ideas or refine one's own original viewpoints. Whatever the case, I was attracted by different notions in Monique's work on the subject of the state that I like to discuss below to provide further food for thought on the debates on state-building.

The centrality of the state

The first notion is that of the renewed interest in the state, or perhaps more precisely defined the continuing importance of the *idea* of the state (Nuijten, 2003: 206). I do not mean here the rediscovery or 'reinvention' of the state due to renewed fears of inter-state conflict or renewed east-west conflagrations as is mentioned frequently these days, but rather the fact that the state is still looked upon as the final remedy for societal or political problems. Monique observes that on the one hand the state is losing influence and sovereignty in a new global order and also due to its own corruption and inefficiency, but on the other continues to play a central role in development programs and in notions of rule and order (Nuijten, 2003: 207). She asserts: 'State power continues to be most important in the imagery of rule and governance … Hence, the idea of the state remains central as the object of fantasies and discourses of order' (2003: 208).

This emphasis on the centrality and necessity of a strong state is also fundamental to the desire to establish, build or strengthen states in a disorderly and dangerous world, as exemplified in global attempts at state and peacebuilding. Though this was partly the result of the post 9/11 conviction that fragile states were unable to control terrorists and deny them a safe haven and of geo-strategical interests, it was also reflecting a deeper belief that a western state conception and *modus operandi* would be beneficial to the societies and populations at stake. As stated by Sisk: 'State-building has become an overarching concept to security and development in fragile states that envisages improvement in governance institutions and processes at the national and local level as a way to channel and manage social conflict away from the battlefield or streets and into regularised processes of non-violent resolution of conflict through professional public administration, elections and parliamentary politics, and through participation and voice of citizens' (2013: 1).

The state as an object of desire and fantasy

In that sense the idea of the state as *the* means to guarantee a safer and better future was a deeply ingrained notion. Such notions became exemplified in the liberal peace and state-building policies that the western world tried to implement in conflict-affected societies from the mid-1990s onwards on an increasing scale. Though this idea that an imported western liberal state model would deliver peace, stability, democracy and development was undoubtedly naïve, arrogant or even deceptive, it reverberated widely and in fact confirmed this imagery and fantasy of rule and order, as observed by Monique.

In a similar vein, Lotz observes that international standards for state-building promoted by globalisation are often appreciated locally and hence receive local support, as they act (or are believed to act) as drivers of progress and offer hope and opportunities (2010: 233). Monique discusses in this connection the state as 'a hope-generating machine' and observed enthusiasm among Mexican *ejidatarios,* for example, about a new president, even if they knew they would be probably deceived by him (2003: 196).

State-building in practice

I do not have to repeat that state-building in reality often showed a completely different face from what it promised to do. Suhrke (2013) describes, for example, how the external state-building intervention in Afghanistan created its own contradictions in the form of external dependency leading to a rentier state, massive corruption and a lack of ownership and legitimacy. Debiel and Lambach point out that state-building projects are designed in western bubbles without any connection to local realities, by external bureaucracies and think tanks removed from grounded realities and operating on the basis of a self-referential logic. These external actors are unable to negotiate the prevailing hybrid political orders or to deal with the socio-political complexities involved (2009: 22 and 26).

State-building dilemmas

Several authors have noted that the state-building exercise is characterised by a number of thorny dilemmas where all imaginable solutions entail risks and costs. With regard to the role of foreign military involvement, Edelstein, for example, asks attention for the duration and footprint dilemmas. The duration dilemma refers to the need for a long presence to produce good results, while this at the same time engenders the 'obsolescence of welcome' where levels of support may dwindle after some time when deaths occur or no tangible results are forthcoming. The footprint dilemma entails the need for an intrusive approach, but that is something that simultaneously undermines local ownership and long-term sustainability (2009: 81). Paris and Sisk also list several contradictions and dilemmas of state-building (2009: 305-309). As there are no easy prescriptions possible for resolving dilemmas, they recommend to better manage them by carrying out a 'dilemma analysis'. This may however hide more fundamental problems related to the nature of state-building rather than how to manage it wisely.

Critiques on state-building

In this respect, several authors go a step further and argue that the whole idea of state-building may not only be (too) difficult or ambitious to implement, but even completely ill-advised. Writers like Richmond and Pogodda state that even though driven by the liberal ideal of a state delivering security and services to its population, state-building practice has effectively failed by design. They argue that 'the neo-liberal states it constructs are subservient to global capital and northern state security interests rather than to human rights, representation or law … fall short of responding comprehensively to economic needs … and fail to provide public services quickly enough to undercut currents of violence or address the root causes of conflict' (2016: 3).

Chandler notes that state-building has nearly become a ubiquitous practice, but that it is addressed in a depoliticised, technical and functionalist language of development or capacity building. In this way the political core of state-building is hidden or denied, hence the title of his book *Empire in Denial*. As explained by Chandler: 'the concept of Empire in Denial attempts to capture the new forms of international regulation of non-Western states and societies: the fact that the new forms of international control attempt to evade responsibility and accountability for the exercise of power' (2006: 10). Apart from eschewing accountability by denying the politics of state-building and rendering the topic technical as illuminated by Chandler, Egnell (2010) introduces the concept of organised hypocrisy as a means to make sense of the inconsistencies and contradictions in the theory and practice of contemporary state-building. Organised hypocrisy can help decouple talk, decisions and actions that are in themselves inconsistent and thereby hide these inconsistencies. Finally, a last critical author, Cramer, likens post-conflict reconstruction efforts to 'The Great Post-Conflict Makeover Fantasy', which he calls 'an enormous experiment in social engineering' (2006: 245-278).

Conclusion

This brings us back to one of Monique's concluding observations about the nature of dreams, fantasies and desires. She states that 'the (self-)image of the western world and the USA as being 'democratic' and 'free' countries organised on the basis of rational-legal principles, which help the rest of the world to become modern and democratic, can be seen as a western fantasy' (2003: 205).

Unfortunately, on the basis of the above we have to conclude that such fantasies are not harmless. They may represent a desire towards hegemonic empire as stated by Chandler, though denied and shrouded in technical development jargon, or hardly less damaging, an arrogant and misplaced hubris of building state institutions in other people's countries and societies. This state of affairs underlines two quotes in the volume by Richmond and Poggoda: 'State-building follows structural power in the international system' (Strange, 1988 quoted in Richmond and Pogodda, 2016: 3) and 'Peacebuilding and state-building are coloured by post-cold war triumphalism, rested on a crude form of capitalism and the erosion of citizens' classical liberal rights and needs' (Richmond and Pogodda, 2016: 4). These quotes and the preceding analysis should not only call for modesty in current state-building efforts, but also a fuller reconsideration of its premises, approaches and practices. Monique's warning to critically look at power, the state, the force fields and organising practices involved is a useful guide in such an exercise.

Georg Frerks / Professor of Conflict Prevention and Conflict Management, Utrecht University, the Netherlands / Professor of International Security Studies, Netherlands Defence Academy / Former Professor of Disaster Studies, Wageningen University, the Netherlands / g.frerks@uu.nl

References

Chandler, D., 2006. Empire in denial. The politics of state-building. Pluto Press. London / Ann Arbor, Mi.

Cramer, C., 2006. Violence in developing countries. War, memory, progress. Indiana University Press, Bloomington and Indianapolis

Debiel, T. and Lambach, D., 2009. How state-building strategies miss local realities. Conflict, Security & Development 21(1): 22-28.

Edelstein, D.M., 2009. Foreign Militaries, sustainable institutions, and postwar statebuilding. In: Paris, R. and Sisk, T. (eds) The dilemmas of statebuilding. Routledge, London and New York. pp. 81-103.

Egnell, R., 2010 .The organised hypocrisy of international state-building. Conflict, Security & Development 10(4): 465-491.

Lotz, C., 2010. International norms in statebuilding: finding a pragmatic approach. Global Governance 16(2): 219-236.

Nuijten, M., 2003. Power, community and the state: the political anthropology of organisation in Mexico. Pluto Press, London / Sterling, Virginia.

Paris, R. and Sisk, T. (eds), 2009. The dilemmas of statebuilding. Routledge, London and New York.

Richmond, O. and Pogodda, S., 2016. Post-liberal peace transitions. Between peace formation and state formation. Edinburgh University Press, Edinburgh.

Sisk, T., 2013. Statebuilding. Polity, Cambridge.

Strange, S., 1988. States and markets. Pinter Publishers, London.

Suhrke, A., 2013. Statebuilding in Afghanistan: a contradictory engagement. Central Asian Survey 32(3): 271-286.

10. HOPE

Joost Jongerden

Mr. Godot told me to tell you he won't come
this evening but surely tomorrow.
Samuel Becket – Waiting for Godot (1955)

Before the law sits a gatekeeper. To this gatekeeper comes a man from the country who asks to gain entry into the law. But the gatekeeper says that he cannot grant him entry at the moment. The man thinks about it and then asks if he will be allowed to come in later on. 'It is possible,' says the gatekeeper, 'but not now.
Franz Kafka – Before the Law (1915)

Introduction

In her anthropology of the political, *Power, Community and the State*, Monique Nuijten (2003) strikingly observed that it is remarkable that at a time when the state appears to lose influence under processes constituting globalisation, anthropologists are showing an increased interest in debates about the state. She went on to say that this can be seen, on the one hand, as a renewed concern with power in an era when political power has shifted away from the national state to a global level of transboundary assemblages (Sassen, 2008), and, on the other hand, as the idea of the state gaining a central position in fantasies of rule (Nuijten, 2003: 1). At the core of her anthropological explorations is the idea of a 'hope-generating machine' (Ibid: 16). This machine is fuelled by the fantasy that everything is possible (Ibid: 174). In this contribution, I will discuss and extend this concept of hope.

Fraudulent Hope

The main product of the hope-generating-machine is the expectation that problems will be solved and hardships overcome. In return, the machine expects patience. It is in this context that bureaucrats offer convincing rationales for endless social justice cases that are opened but never closed, expectations raised for a different future that go ever unfulfilled. The gaps between the now, the real existing conditions, and future, the promise of a not-yet, are stuffed and brimming over with hope. This hope, however, has a particular set of features.

First, it is a messianic hope, since it is based in a belief that the bureaucracy, mediating between the disadvantaged and an abstract higher power, will bring salvation, eventually. Second, it is a hope that pacifies, since as long as the case is pending, one can only wait. Third, it is a hope which subjugates, since the hope that justice will be done keeps people obedient. Fourth it is dynamic, since with every election of a new president, and likely its miniature versions at provincial and local level, new fuel for hope is injected in the machine.

Focusing on the lives of ejidatarios and landless families in rural Mexico, the term 'hope-generating machine' captured well the bureaucratic practices that produce and proliferate expectations for improvement and progress – although maybe 'hope-generating *rhizome*' would be a better term, since the controlling bureaucracy is not centrally commanded but composed of thousands of independent practices. Machine or rhizome, the hope-generator is not there to deliver but to produce a forever not-yet (Nuijten, 2003: 227). As such, the hope produced could be defined as a 'fraudulent' hope (Bloch, 1995) or 'naïve' hope (Freire, 2014).

However, Monique Nuijten warns, the machine that produces this fraudulent, naïve hope has another production line: violence. Hope that is never fulfilled is likely at some point to turn into despair and anger. When peasants and villagers are tired of 'waiting for Godot' and no longer willing or able to act out the phantasm of the hope-generating-machine, they can resign themselves to their fate or else act upon their desires. In the case of the latter, when people stand up to fight for their rights, for a better life, for justice and equality, the hope-generating machine starts to spit violence:

> *At the point when the creation of hopes by the bureaucracy is no longer sufficiently effective, and when peasants no longer want to engage in these incredible fantasies, the latter are confronted with threats, murders and other forms of violence. At that point, the peasants are 'pushed out of' or, in other words, 'excluded' from the regular system.*
> Nuijten, 2004: 227

Hope and the Political

Crucially, hope is a concept for understanding the political, the social relations of power. Monique Nuijten conceptualised these relations of power as a force field, a dynamic ordering in which relations of domination are constructed, identities forged, contention occurs and resistance may develop (Nuijten, 2003: 12). Thus it was that when scrolling through a critical lexicon of political concepts recently – a book that the editors, Bernstein *et al.* (2018), would have as reviving our political vocabulary, or at least aiming in that direction – I could not supress a feeling of disappointment. The lexicon included chapters entitled 'Power', 'Exploitation, 'Identity' and suchlike – but not 'Hope'. Yet hope would surely be an appropriate not to say excellent entry in a book attempting to engage with the question of 'the political'. The question of the political – or more precisely, the configuration of relations of power – is precisely the issue that Monique Nuijten engages with when exploring the concept of hope, and its companion, spewed out in excess by the hope-generating machine, violence.

Though somehow disappointed, one should not be surprised. We ought to acknowledge that few scholars have concerned themselves with the concept of hope like Monique Nuijten. Three who did, however, might be mentioned here. Ernst Bloch was one, a fringe member of the *Frankfurter Schule* who examined 'how hope functions in the world as a real latent force' (Zimmermann, 2013: 4). Bloch's three-volume compendium *The Principle of Hope* (1995) contrasts sharply with the growing pessimism of the influential and iconic leading members of the *Frankfurter Schule*, Theodor Adorno and Max Horkheimer (Horkheimer and Adorno, 2002; Horkheimer, 2004). For Bloch, hope is not only linked to the expectation that something may happen, God willing, but is first and foremost rooted in our daily lives and the social struggle (Zimmermann, 2013: 8). Hope is action. For Bloch, therefore, hope not only anticipates a better world in the indefinite future but also involves action and engagement now, in the present (Giroux, 2004). Although hope is an expectation, it is not brought to our door by a delivery service. Hope is enacted.

Another scholar who has considered hope is Paolo Freire. From his work, one can makes a distinction between the type of naive hope produced by the machine and a critical hope (Grain and Lund, 2016):

> *Without a minimum of hope, we cannot so much as start the struggle. But without the struggle, hope, as an ontological need, dissipates, loses its bearings, and turns into hopelessness. And hopelessness can become tragic despair.*
> Freire, 2007: 9

Thus, Freire characterises the hope that awaits its realisation from an external power, divine or sovereign as naive and that which is rooted in social struggle and questioning as critical. This brings me to the third thinker in my short list, someone whose work I increasingly came to appreciate when I started to teach on a course that Monique Nuijten coordinated and which was organised around one of his books: Zygmunt Bauman.

Critical Hope

Referred to as the 'sociologist of misery' (Dawson, 2012: 555), the work of Bauman is usually not associated with hope. And it's true that Bauman analyses modernity, paraphrasing Monique Nuijten, as a 'human-waste-generating machine'. He certainly does turn our gaze to the ways in which modernity's obsession with order-making continuously produces excess, populations who are considered to spoil the order and so need to be removed (Bauman, 2003). In *Modernity and the Holocaust*, Bauman (1989) argued that the holocaust was not an interruption of modernity, a backsliding into barbarity, but that it was actually modernity itself that constituted the most decisive factor in the horror. So far, so bleak. In lectures and interviews, moreover, Bauman similarly failed to cheer up his audience. At a 2004 conference in Leiden, for example, he opened his talk with the words 'Another day I had a nightmare' (Bauman 2004). And when asked about his personal life in an interview for Al-Jazeera, he answered quoting Goethe that he had a happy life but couldn't remember a single happy week (Williams, 2017).

Still, several scholars have referred to Bauman's work as a 'sociology of hope' (Davis, 2011; Dawson, 2012: 555; Giroux, 2004; Zimmermann, 2013). And actually, Bauman too, refers to his thought with the term 'hope'. When asked for Czech television why he was so pessimistic, Bauman's answer began

> *I'm neither a pessimist nor an optimist, because I believe that the difference between them is that the optimist says this world, our world, is the best possible world, and the pessimist suspects that this optimist is right. That's the only difference between them.*

But then he added

> *I have a third category, and that is the category of hope.*
> Wise Society, 2011

Bauman's hope is based on the conviction that the future does not have to be an extension of the present; a different world really is possible. It is likely he would regard the procrastination of the hope-generating machine as yet another dimension of social inequality:

> *The drama of power hierarchy is daily restaged (…) in innumerable entrance lobbies and waiting rooms (…) Position in the hierarchy is measured by skill (or ineptitude) in reducing or cutting out completely the timespan separating a want from its fulfilment. Climbing the social hierarchy is measured by rises in the ability to have what one wants (whatever it may be) now – without delay.*
> Bauman, 2003

Therefore, Bauman's hope is not identified with waiting, but with action. It is based on a culture of questioning with the purpose to

> *(…) keep the forever unexhausted and unfulfilled human potential open, fighting back all attempts to foreclose and pre-empt the further unravelling of human possibilities, prodding human society to go on questioning itself and preventing that questioning from ever stalling or being declared finished.*
> ibid.

In their own ways, both Bloch (1995) and Bauman (2010) believe that hope cannot be removed from this world, even in the darkest times, because it represents the *not yet* in the sense of a possibility that may become real if we work for it (Giroux, 2004: 39). This is a not the naive hope of a waiting for something to happen but a critical hope that emerges from doing, questioning and learning. It is not the messianic hope produced by the machine, which turns violent when its hope turns phantom, but one that is anticipatory and mobilising at the same time, because its desire for a better future is grounded in daily practices and present struggles. I believe that in her later work, Monique Nuijten moves from discussing the hope produced by the hope-generating machine to the anticipatory and mobilising hope in the not-yet of grassroots activism (Nuijten, 2015: 481, 483).

In my own work on Kurdistan and Turkey (Jongerden, 2018, 2019; Öztürk *et al.*, 2018), I have focused on practices through which people try to 'change the world that is changing them' (Berman, 1982). Thus, we observe the peasant household taking advantage of job-opportunities in the city to compensate for deteriorating prices for their agricultural produce or competition, so-called, from the world market, to city-dwellers maintaining backward links with their rural hinterland as a support in the face of income insecurity in the city. On looking closely, we see these actions as more than just individual coping mechanisms. We see the formation of new collective identities around extended families, hometown associations and related forms of solidarity. And we see forms of informal and rebel governance.

The many ways in which people – individually and collectively – organise themselves to create a better future in the here and now is a major focus of my research. Particularly in the context of progress and (post-)modernism, I refer to it as a 'Do-It-Yourself development'. A key question revolves around the practices through which people themselves create and maintain a *liveable life* under conditions of inequality and insecurity. How this hope for a better life is created is well captured by the Kurdistan Workers Party's (PKK) slogan '*Berxwedan Jîyane*': Resistance is Life. It is in the struggle itself that hope is produced.

Hope, finally

Thus, staring from Monique Nuijten's concept of the hope-generating-machine, two forms of hope may be distinguished. The first is a messianic hope as primary product of the hope-generating-machine, which pacifies, subjugates and procrastinates. This hope is based on the belief that its fulfilment comes from above and is disconnected from social struggle. The second is a critical hope. This is a hope based on doing, questioning and learning, a hope grounded in social practice.

When the machine that generates hope jams, it may deliver its other service, violence; but what about the critical hope, the hope generated in struggle? What happens when this hope is disappointed? When Bloch was asked if this hope could be disappointed, he replied

Even a well-founded hope can be disappointed, otherwise it would not be hope. In fact, hope never guarantees anything. It is characteristically daring and points openly to possibilities that in part depend on chance for their fulfilment.
Zipes, 2020

Bloch also noted, however, that

Hope can learn and become smarter through damaging experiences.
ibid.

Indeed, hope may involve 'learning from defeat' (Jongerden, 2019) and be an 'action-oriented approach to contemporary despair' (Grain and Lund, 2016). This is the hope of people, who, acting in concert, 'challenge the status quo in a quest for and enactment of new utopias' (Nuijten, 2015: 483).

Joost Jongerden / Associate Professor, Rural Sociology Group, Wageningen University, the Netherlands / joost.jongerden@wur.nl

References

Bauman, Z., 1989. Modernity and the holocaust. Cambridge, Polity Press.

Bauman, Z., 2003. Wasted lives, modernity and its outcasts. Cambridge, Polity Press.

Bauman, Z., 2004. Liquid modernity: lecture on the ANSE conference. ANSE Conference on Value Dilemmas as a Challenge in the Practice and Concepts of Supervision and Coaching. Leiden, the Netherlands.

Bauman, Z., 2010. Towards a critical sociology: an essay on common sense and emancipation. London, Routledge.

Berman, M., 1982. All that is solid melts into air. New York, Penguin Books.

Bernstein, J.M., Ophir, A. and Stoller, A.L., 2018. Political concepts: a critical lexicon. New York, Fordham University Press.

Bloch, E., 1995. The principle of hope (Vols. 1-3). Boston, MIT Press.

Davis, M., 2011. Bauman's compass: navigating the current interregnum. Acta Sociologica 54(2): 183-194.

Dawson, M., 2012. Optimism and agency in the sociology of Zygmunt Bauman. European Journal of Social Theory 15(4): 555-570.

Freire, P., 2007. Pedagogy of hope. New York, Continuum.

Freire, P., 2014. Pedagogy of hope: reliving pedagogy of the oppressed. London, Bloomsbury.

Giroux, H.A., 2004. When hope is subversive. Tikkun 19(6): 38-41.

Grain, K.M. and Lund, D.E., 2016. The social justice turn: cultivating 'critical hope' in an age of despair. Michigan Journal of Community Service Learning Fall 2016: 45-59.

Horkheimer, M., 2004. Eclipse of reason. London, Continuum.

Horkheimer, M. and Adorno, T., 2002. Dialectic of enlightenment: philosophical fragments. Stanford, Stanford University Pres.

Jongerden, J., 2018. Living structures. Methodological approaches in Kurdish studies: theoretical and practical insights from the field. Boulder, Lexington Books: pp. 21-33.

Jongerden, J., 2019. 'Learning from defeat: development and contestation of the 'new paradigm' within the Kurdistan Workers' Party (PKK).' Kurdish Sudies 7(1): 72-92.

Nuijten, M., 2003. Power, community and the state. London, Pluto Press.

Nuijten, M., 2004. Between fear and fantasy governmentality and the working of power in Mexico. Critique of Anthropology 24(2): 209-230.

Nuijten, M., 2015. Political activism and the PAH (Platform of Mortgage Victims) in Spain: the significance of Franz von Benda-Beckmann's work for the study of power, political agency and legal pluralism at the grassroots. The Journal of Legal Pluralism and Unofficial Law 47(3): 476-492.

Öztürk, M., Jongerden, J. and Hilton, A., 2018. The (re)production of the new peasantry in Turkey. Rural Studies.

Sassen, S. (2008). Territory, authority, rights from medieval to global assemblages. Princeton, Princeton University Press.

Williams, F., 2017. This is not an obituary. The Bauman Isntitute. Available at: https://baumaninstitute.leeds.ac.uk/this-is-not-an-obituary/.

Wise Society, 2011. Zygmunt Bauman: quest'economia ci consuma [video]. Vimeo. Available at: https://vimeo.com/31550442.

Zimmermann, R.E., 2013. Transforming utopian into metopian systems: Bloch's principle of hope revisited. In: Thompson, P. and Žižek, S. (eds) The privitization of hope: Ernst Bloch and the future of Utopia. Durham, Duke University Press, pp. 246-268.

Zipes, J., 2020. Ernst Bloch and the Philosophy of Hope. Tribune Available at: https://www.tribunemag.co.uk/2020/04/ernst-bloch-and-the-philosophy-of-hope.

11. THE PACIFIC 'HOPE-GENERATING MACHINE': NEGOTIATING DEVELOPMENT IN SOLOMON ISLANDS

Stephanie Ketterer Hobbis

Researchers in, and of, the Pacific Islands have long struggled with understanding why Pacific Islanders perpetually seek to attract the attention of national governments and international donors to facilitate large-scale development projects, just to be disappointed over and over again (e.g. see Oppermann, 2015; Tammisto, 2016). I faced the same challenge: During my fieldwork in 2014, the Lau-speakers of Malaita Province, Solomon Islands, were eagerly participating in a three-day state-sponsored event aimed at reconciling a conflict about ownership of land and land-covered-by-sea and at facilitating the construction of a tuna cannery, international sea- and airport and township.[1] Villagers' enthusiasm for this event seemed by no means diminished by the fact that conflicts had always reemerged shortly after similar events in the past. On the contrary, for nearly two months, the residents of various villages enthusiastically reoriented much of their daily activities to prepare for this event. They practiced dance routines, procured necessary foods, and worked out the logistics that would ensure that even relatives living in the furthest corners of the country would be able to attend.

[1] Following requests from my Lau interlocutors, I will not discuss the particularities of the conflict that led to the reconciliation event.

Without Monique Nuijten's work on state bureaucracies and development as 'hope-generating machine' that 'generates enjoyments, pleasures, fears and expectations' (Nuijten, 2003: 16), I would not have been able to make sense of this enthusiasm and villagers' commitment to yet another reconciliation event. Monique, thus, became in many ways the inspiration for my own work, an inspiration for a cornerstone of my PhD thesis, and my reflections on Lau engagements with 'development' ever since. Pacific bureaucracies also '[offer] endless openings' and their officials also seem nearly 'always willing to initiate procedures' (Nuijten, 2003: 16), especially when it comes to initiatives that aim to solve land conflicts for the specific purpose of large-scale development projects. Hence, when a previous 'hope-generating event' for the cannery had failed, a Memorandum of Understanding between the Malaita Provincial Government and landowning clans in 2010, Solomon Islands bureaucrats did not abandon the project. On the contrary, when re-approached by the conflicting landowning clans with a new proposal for reconciliation state officials not only supported another initiative but provided even more funds than they had for the previous one. The event that I observed would be the largest and most lavish that had happened yet, including an extravagant feast for which over fifty pigs were to be slaughtered.

As Monique so succinctly pointed out, 'people are never naive,' but still, 'during certain periods they can become inspired and enthusiastic about new programmes and new openings that are offered to them' (2003: 197). I saw this enthusiasm reflected in my interlocutors' preparations for the festivities associated with the reconciliation event and in the few critical perspectives on the event brought up during these preparations. Disappointment was, however, just around the corner and shortly after, the tone switched to a more critical assessment of the event and what it may, or may not achieve. In fact, there was almost immediately consensus that the conflict had not been adequately solved and that more reconciliations would be necessary in the future if the cannery project was to move ahead. Many of my interlocutors were disappointed, but many were also convinced that another opportunity would present itself. Some even expressed instant excitement about the possibilities for an even more elaborate event in the near future, with more pigs and even better performances. The 'hope-generating machine' kept on moving, and no one that I talked to seemed to have any significant doubt about their ability to convince bureaucrats to support their reconciliation efforts, if necessary, over and over again. I am eternally grateful for Monique's inspiration to uncover, and perpetually trace, this machine as it grinds on in a 'never-ending cycle of high expectations followed by disillusion and ironic laughter' (Nuijten, 2003: 197), and shortly after, hope again.

Stephanie Ketterer Hobbis / Assistant Professor, Sociology of Development and Change Group, Wageningen University, the Netherlands / stephanie.hobbis@wur.nl

References

Nuijten, M., 2003. Power, community and the state: the political anthropology of organization in Mexico. London: Pluto Press.
Oppermann, T.C., 2015. Fake it until you make it: Searching for mimesis in Buka village politics. Oceania 85(2): 199-218.
Tammisto, T., 2016. Enacting the absent state: State-formation on the oil-palm frontier of Pomio (Papua New Guinea). Paideuma 62: 51-68.

12. MONIQUE NUIJTEN AND THE 'HOPE-GENERATING MACHINE' IN THE TIME OF COVID-19

David Mosse

Over the past several months of this year of Covid-19 (2020), citizens in many countries have daily been spoken to by their political leaders, standing behind podiums, flanked by scientific experts, using graphs and graphics to chart infection rates, hospitalisations and deaths; and to convey instructions on private and public behaviour in a manner quite unimaginable until this pandemic spread.

The rising curves of affliction and appeals to citizens to act to protect healthcare institutions and the social fabric take us into territory that falls well beyond the displays of coherence and rationality that characterise what we understand by government. The limits of the state's foundational power to make live and let die are made apparent by the uncertainties produced by faltering efforts to govern the virus (Smith, 2020). Amidst the variety of concepts of state and government repurposed to make sense of the political responses to the Covid-19 pandemic (e.g. Hannah *et al.*, 2020), Monique Nuijten's idea of state bureaucracy as a 'hope-generating machine' stands out for its capacity to grasp the present moment. Under conditions of the pandemic, the state draws in and feeds fears, expectations and hopes, where coherence cannot be had (2003: 16). Whatever has to be engineered by way of changed behaviour, curtailed freedoms and economic losses depends, we increasingly realise, upon the generation of twinned fear and hope.

In its hope-generating mode, Nuijten says (of the Mexican situation she examined in quite other circumstances),

> *[t]he bureaucracy offers endless openings, and [that] officials are always willing to initiate procedures. The bureaucracy as a hope-generating machine gives the message that everything is possible... The bureaucracy never says no and creates great expectations. On the other hand, many promises are never fulfilled.*
> Nuijten, 2004a: 211

We are in the midst of a hope-generating moment, where politicians announce that 'whatever it takes' will be done, ambition is limitless (for now), new procedures and radical forward-looking are the character of governments in which accountability for decisions and implementation choices is deferred to the future. The conditions of crisis bring to the fore that aspect of government which, as Nuijten explains, depends upon the deployment of myths and fantasies.

What is important here is to have a way of looking at the processes and practice of state and citizenship which Nuijten's work brings, and that might enable us to better trace and understand what is happening around us, both from the viewpoint of state processes, and of altered forms of citizen 'self-making' amidst Covid (Lazar and Nuijten, 2013).

The idea of the 'hope-generating machine' is among a set of concepts born of Monique Nuijten's deep study of state, resources struggles and local institutions affecting a peasant community (*ejido*) in western Mexico in the 1990s, which have been applied more widely in international development, and which now prove to be salient in this moment of governing uncertainty. Monique's work on organisations, for example, is relevant now it is clear that the response that is needed to the pandemic far-exceeds that falling within the provenance of any given organisation. We have to rethink the nature of organisations, the boundaries that define them and their formal rules of operation, and recognise again the importance of organisational narratives (Nuijten, 2005); issues that have been the focus of much of Nuijten's work.

We can now learn from her insight – derived from the study of power and institutions for natural resources management (2005) – that organising does not require organisations, that informal settings and personal networks of influential actors or brokers are a better focus in order to see how coordinated or collective action comes about. Many may now also feel the gap between people and the state; the gap (in Mexico) that drew Monique's attention to the realm of brokerage and organising practices between communities and the state. How people accommodate to the presence of Covid-19 will very much depend upon organising practices 'from below' (Nuijten, 2003: 198-199). And yet we often see precisely what Monique Nuijten anticipated state interventions (now in response to Covid-19) amplifying inequalities through the brokered provision of services and resources and forms of connection and 'forcefields of power' which diminish the capacity of the most disadvantaged to negotiate access or protection of welfare or livelihoods (ibid). She also points to the power of discursive systems which appropriate ideas from one context (maybe theory) for use in another (policy intervention) (2004b), noticing the fast pace of shifts in language and policy modelling, which are only amplified in current circumstances.

My point here is that Monique Nuijten exemplifies the capacity of anthropology to discover, through careful examination of social and institutional processes in one context, by the means of ethnographic immersion in these processes, characteristics of human organisation that have far wider relevance. This is what made her contribution so important to the creative anthropology of development work of the early 2000s, partly fostered through the EIDOS network and its conferences, workshops and publications. It is also what makes her ideas relevant today when processes of government and organisation are challenged and are adapting in unprecedented circumstances. And from this body of work we acquire key concepts and a methodology to investigate the agency of citizens, compliance with restrictions, making claims and testing the government, subjecting it to scepticism, and examining the 'dividing effects' of state/citizen processes (2003: 198).

The relevance of this work in the present moment is testament to its power and flexibility and the capacity of its insights to transcend the context of their production. Indeed, the mark of truly valued anthropological thought is that it offers ways of thinking about entirely new circumstances such as we face now with the Covid pandemic. We are, and will continue to be, grateful for the intellectual and personal generosity of Monique Nuijten's contributions.

David Mosse / Professor of Social Anthropology, SOAS University of London, United Kingdom / dm21@soas.ac.uk

References

Hannah, M.G., Hutta, J.S. and Schemann, C., 2020. Thinking through Covid-19 responses with foucault – an initial overview. Antipode Online. 5 May 2020 https://antipodeonline.org/2020/05/05/thinking-through-covid-19-responses-with-foucault/.

Lazar, S, and Nuijten, M., 2013. Citizenship, the self, and political agency. Critique of Anthropology 33(1): 3-7.

Nuijten, M., 2003. Power, community and the state: the political anthropology of organisation in Mexico. London: Pluto Press.

Nuijten, M., 2004a. Between fear and fantasy: governmentality and the working of power in Mexico. Critique of Anthropology 24(2): 209-230.

Nuijten, M., 2004b. The hope-generating machine and the positionality of the anthropologist. In: Salemink, O., Van Harskamp, A. and Giri, A.K. (eds) The development of religion/the religion of development. Delft: Eburon.

Nuijten, M., 2005. Power in practice: a force field approach to natural resource management. Journal of Transdisciplinary Environmental Studies 4(2): 3-14.

Smith, R. 2020. What type of governmentality is this? Or, how do we govern unknowns. Somatosphere, May 26. Available at: http://somatosphere.net/2020/governmentality-covid19.html/.

13. FORCE FIELDS SURROUNDING TROPHY HUNTING IN NAMIBIA: GENERATING HOPE WITH ECONOMIC BENEFITS FOR THE SAN OF NAMIBIA?

Stasja Koot

Introduction

Recent years have shown an increase in the, often heated, debate on trophy hunting, with some important developments taking place in southern Africa. In 2012, pictures of King Juan Carlos of Spain emerged in which he posed in front of his trophies, an elephant and two African buffaloes. As a result of the public outcry that followed, the King was dismissed as the honorary president of WWF Spain, which is ironic when realising that various WWF offices in southern Africa support trophy hunting in the name of conservation and development. Other important developments were the ban on trophy hunting that was introduced in Botswana in 2014 (and lifted again in 2019), a very controversial hunt of a black rhino in Namibia (for US$350,000) in 2015 and the infamous illegal hunt of Cecil the Lion in Zimbabwe in that same year. In this essay, I focus on trophy hunting in Namibia, analysing it through two concepts developed by my friend and colleague Monique Nuijten: 'hope-generating machine' (Nuijten, 2003) and 'force fields' (Nuijten, 2005). As I show, both concepts help to disclose important elements in the power dynamics of trophy hunting.

Generating hope through trophy hunting

Proponents of trophy hunting argue that it is good for conservation *and* for the development of marginalised people (Angula *et al.*, 2018; Naidoo *et al.*, 2016). In Namibia, a flagship country for so-called 'community-based natural resource management' (CBNRM), trophy hunting plays a crucial role in this programme; in CBNRM, local groups are targeted to contribute to conservation in return for economic and material benefits (e.g. tourism jobs and the hunted meat) (Koot and Van Beek, 2017; Naidoo *et al.*, 2016; Sullivan, 2002). However, to keep CBNRM financially healthy, a continuous stream of income is required, and trophy hunting provides for such large revenues because the amounts that hunters pay can be enormous. In the Namibian Nyae Nyae Conservancy, for instance, the price for a 14-day hunt of one elephant is US$ 80,000 (Paterniti, 2017).

Conservationists, hunting operators and the Namibian Ministry of Environment and Tourism (MET) argue that the hunting industry is very important for the development of marginalised rural populations. By continually articulating this idea publicly (e.g. in the media) and at community meetings, they create hope for marginalised groups involved in the CBNRM program, thus creating what Nuijten (2003: 16) called a 'hope-generating capacity'. However, according to Economists at Large (EAL, 2013), trophy hunting contributes only 0,27% to the Namibian GDP, and most of these revenues go to hunting operators, airlines, governments and tourism facilities. So how serious should we take the argument that 'economic benefits' – that create this hope in the first place – truly reach the rural poor of Namibia? This purely economic argument on which the hope is based, is problematic for various reasons; it masks differences within communities that are presented as if they are static, homogeneous entities *and* it masks power relations between segments of these communities and outsiders (e.g. NGOs, hunting operators, donors and government officials), crucial relations that Nuijten described crystal clear as 'force fields'. As she explains, power is often articulated as static, as something that people 'possess', but such a notion 'ignores the fundamental fact that power is always 'relational' and the result of the working of multiple, intertwined institutions' (Nuijten, 2005, 1). Of course, economic benefits exist, but by focusing only on how these prove the success of trophy hunting they also function as a continuous generator of hope within this force field (Nuijten, 2003), while other dynamics are covered up, especially those in the social and human domain (Koot, 2019). Moreover, it supports the idea that 'local' people do not yet understand how to do 'proper' conservation and therefore need to be educated (MacDonald, 2005). This strongly resembles colonial structures and power relations, based on paternalistic ideas about moral edification, thus again showing the importance of looking at such relations as fluid force fields: 'more structural forms of power relations, which are shaped around the access to and use of specific resources' (Nuijten, 2005: 2).

Force fields around the San of Namibia

An interesting group in this regard is the San of Namibia. As 'former' hunter-gatherers some San groups are today involved in CBNRM initiatives in which trophy hunting plays a crucial role. Take, for example, the Khwe San who live in Bwabwata National Park. When some Khwe tried to establish relationships with wealthy, white hunting operators, this was considered 'bribery' by the MET and a local NGO working on CBNRM, because officially contact with hunters is not allowed when the selection process for a tender is not yet finished. However, 'on the ground' such negotiations often take place informally (Koot, 2019; Koot *et al.*, 2016) in fluid force fields between a variety of actors (cf. Nuijten, 2005). Furthermore, the question remains if hunting operators are indeed the right people to be involved in 'development'; a hunter in Bwabwata recently explained that the 'lazy' Khwe employees are mostly a hindrance to his business and he has no interest in community development (anonymous personal communication). In another example, the Ju/'hoansi San of the Nyae Nyae Conservancy showed how the group that worked with one particular hunting operator felt stressed, suppressed and humiliated by the operator, while receiving very low wages. When I asked them why they would not simply leave the job, they mentioned the few livelihood opportunities in the Nyae Nyae Conservancy. Indeed, partly due to the CBNRM programme, their possibilities to start other livelihoods (e.g. based on agriculture) are very limited. Moreover, the distribution of meat has always been a problem because of the wide dispersal of the settlements in Nyae Nyae. According to the Ju/'hoansi labourers, WWF played a crucial role in choosing this particular hunting operator because he would pay the highest bid for the tender (and thus create the largest income for the CBNRM programme). In another example, however, Ju/'hoansi labourers were much more positive about a (subcontracting) hunting operator (Koot, 2019; Koot and Van Beek, 2017). Having said that, within the force field of neo-colonial relations between the San and trophy hunting operators, generating hope continues to be a red thread (cf. Nuijten, 2003: 2005) publicly, and in preparatory community meetings: the public discourses (by NGOs, donors, hunting operators) remain focused on the creation of economic benefits, i.e. jobs and meat. However, once a hunter has been chosen, neo-colonial force fields are uncovered 'on the ground'.

Jobs can thus be perceived as good *or* ill, but to look at them and simply call them 'benefits' masks structural issues in labour, potentially including exploitation. Therefore, I argue for an expansion of the trophy hunting debate beyond economic 'benefits' (Koot, 2019) and to include analyses of power, for which the concept of force fields (Nuijten, 2005) can be crucial. By doing so, local perceptions, meanings, multiple experiences and power relations are addressed, and the larger human domain is taken into account (of which the 'economy' is only a part). In fact, reducing 'development' only to 'economic benefits' creates a simplistic image of a very complex reality. This is of a more general importance in environmental and development issues globally. Together, the concepts of 'hope-generating machine' (Nuijten, 2003) and 'force fields' (Nuijten, 2005) have thus proven their value far beyond Nuijten's original (field)work, showing the value of her conceptual contributions to the social sciences more broadly.

Personal note

I wish to end this essay by thanking Monique wholeheartedly for being an inspiring and friendly colleague. When I arrived in Wageningen in 2015 – quite blue I admit – I remember that, together with Elisabet Rasch, Rob Fletcher, and Bram Büscher, we wrote a document on what it means to be a 'good academic'. Many of the important values that you were so well aware of (and that are now core in the document) have been a real inspiration, and continue to be so to this day.

Stasja Koot / Assistant Professor, Sociology of Development and Change Group, Wageningen University, the Netherlands / Senior Research Fellow Department of Geography, Environmental Management & Energy Studies, University of Johannesburg, South Africa / kootwork@gmail.com

References

Angula, H., Stuart-Hill, G., Ward, D., Matongo, G., Diggle, R. and Naidoo, R., 2018. Local perceptions of trophy hunting on communal lands in Namibia. Biological Conservation 218: 26-31. https://doi.org/10.1016/j.biocon.2017.11.033

Economists at Large (EAL), 2013. The 200 million question: how much does trophy hunting really contribute to African communities? Report for the African Lion Coalition, prepared by Economists at Large. Available at: https://www.ecolarge.com/wp-content/uploads/2013/06/Ecolarge-2013-200m-question-FINAL-lowres.pdf.

Koot, S., 2019. The limits of economic benefits: adding social affordances to the analysis of trophy hunting of the Khwe and Ju/'hoansi in Namibian community-based natural resource management. Society & Natural Resources 32(4): 417-33. https://doi.org/10.1080/08941920.2018.1550227

Koot, S., Van Beek, W. and Diemer, J., 2016. The Khwe of Namibia: foragers between game, tourism and politics. Anthropos 111(2): 497-511. https://doi.org/10.5771/0257-9774-2016-2-497

Koot, S., and Van Beek, W., 2017. Ju/'hoansi lodging in a Namibian conservancy: CBNRM, tourism and increasing domination. Conservation and Society 15(2): 136-46. https://doi.org/10.4103/cs.cs_15_30

MacDonald, K., 2005. Global hunting grounds: power, scale and ecology in the negotiation of conservation. Cultural Geographies 2005(12): 259-91. https://doi.org/10.1191/1474474005eu330oa

Naidoo, R., Weaver, C., Diggle, R., Matongo, G., Stuart-Hill, G. and Thouless, C., 2016. Complimentary benefits of tourism and hunting to communal conservancies in Namibia. Conservation Biology 30(3): 628-38. https://doi.org/10.1111/cobi.12643

Nuijten, M., 2003. Power, community and the state: the political anthropology of organisation in Mexico. London: Pluto Press.

Nuijten, M., 2005. Power in practice: a force field approach to natural resource management. The Journal of Transdisciplinary Environmental Studies 4(2): 3-14.

Paterniti, M., 2017. Trophy hunting: should we kill animals to save them? In National Geographic. Available at: https://www.nationalgeographic.com/magazine/2017/10/trophy-hunting-killing-saving-animals/.

Sullivan, S., 2002. How sustainable is the communalizing discourse of 'new' conservation? The masking of difference, inequality and aspiration in the fledgling 'conservancies' of Namibia. In: Chatty, D. and Colchester, M. (eds) Conservation and mobile indigenous peoples: displacement, forced settlement, and sustainable development. Oxford: Berghahn Press, pp. 158-187.

14. ABOUT FORCE FIELDS, FOUCAULT, AND GOVERNANCE: MONIQUE NUIJTEN'S CONTRIBUTION TO LEGAL ANTHROPOLOGY

Han van Dijk

Introduction

In this contribution I want to highlight Monique Nuijten's contribution to legal anthropology. In the course of her career she has shown a keen interest in topics that are at the core of legal anthropology. Throughout her work, she developed a strong focus on land, natural resources and the workings of politics. Partly, her engagement was born out of a long-standing friendship with Franz and Keebet von Benda Beckmann, the leading scholars in legal anthropology in the Netherlands. Franz as Professor at the Agrarian Law group, was also a colleague at Wageningen University. Franz and Keebet also visited her when she was in the field in Mexico. But foremost Monique's interest in legal anthropology shows her long-term engagement with the plight of Mexican peasant communities whose struggles for more secure access to land are in the centre of her work on land and natural resources.

This engagement is visible in her academic work and throughout her entire career. To me it would even be justified to say (though probably Monique would deny it herself) that her later engagement with social movements has its origins in the same interest in plurality and in the diversity (plurality) of ways in which people try to improve their lives. In preparing this short essay I went back to the times when I read Monique's work and how it helped me to better understand the land and natural resource issues I was studying. In the following I will discuss what I liked most, what insights they gave me, and the important lessons I learned from Monique's work.

Connections with legal anthropology

In her thesis (Nuijten, 1998), the book she made out of her thesis *Power, Community and The State: The Political Anthropology of Organisation in Mexico* (Nuijten, 2003) and a couple of subsequent articles (Nuijten, 2004a, 2005), central concerns of legal anthropology in land natural resource management and power are focal points of her work. Another legal anthropology concern she addresses, is the engagement with the state as a central actor in the fields of power in which her informants operate. Her book and thesis are characterised by rich and very detailed ethnography in which land as the central territorial element of the *ejido* is put centre stage. This is where Monique and I have many points in common. Though we worked at the time in very different settings (Mali and Mexico, pastoralists and peasants), our research participants have a lot of similar struggles centring on land, natural resources and property.

Another similarity is our interest in power and power relations around resources, the field where political and legal anthropology meet. Natural resource management and land tenure and property relations is not about the relations of people with resources, it is about relations between people in relation to these commodities. These relations are mediated by rules, laws if you like, institutions, political relations, efforts to structure decisions about these objects (governance), and claims in the form of discourses, documents, policies and speech acts.

Perhaps the most important thing I learned from Monique is how to look at power in natural resource management. Having started as an engineer (and anthropologist), studying natural resource management, and being bred as a researcher by legal anthropologists, I focused on legal pluralism, rules and institutions and even resource variability. In the course of doing this, I also developed an interest in power, but probably I had a naïve conception of power. The insight that power is relational is key (for me) in moving legal anthropology beyond the analysis of the interaction between actors and plural legal systems. It connects the analysis to Foucauldian notions of governmentality and to the importance of subjects in this Foucauldian vision on power as relational. It connects the analysis of law and legal issues to the level of politics and political relations.

Monique beautifully formulates this with her notion of 'force field' as the locus of relations of force. It teaches us that power is produced between two poles and may also be discontinuous and not the same in all places. Force fields 'are always in flux, it is not possible to 'freeze' them in terms of social or territorial boundaries. Yet they can have a certain stability for a period of time' (Nuijten, 2005: 2). It means that just as power, force fields in which power is produced require efforts of those who want to exert power. Just as political relations have to be produced, maintained, practiced and created, power and the force field in which they operate have to be produced by those people who constitute the political. Without effort on the part of the beholders of power and the preparedness to obey of those subjected, power does not exist.

This brings Monique to another point: in analysing a land conflict in Mexico. In this analysis she formulates an important critique on the governmentality approach and an important addition. Her critique is that focusing on institutional practices with standardised administrative techniques does not make much sense in the Mexican context to understand the workings of power (and in many other contexts as well where clientelism and informal politics play and important role). Instead Monique argued, and this is her addition to the governmentality approach, that we should focus on decentred practices of government and the ways in which subjects engage with these decentralised practices through self-regulation and fantasies of the state and its power symbolised by a 'ritualised world of bureaucratic practices' (Nuijten, 2004: 227).

Monique arrives at these insights, because as a good field worker she mainly focused on the people in her fieldwork site, letting herself be surprised by the apparent incoherence in the strategies of those subjected to power and the almost chaotic way in which they tried to access the state to get a solution for their land claims. Her wonder about the phantasies of state power and her analysis of these makes us understand that the legal is also political, and that natural resource management is not a rational process in which rational actors try to pursue their best interests. Actors have all kind of other considerations, such as the maintenance of clientelist relations, internal group cohesion or the internal conflicts of interest.

There is also an important methodological lesson I learned from Monique's work. In a book chapter, entitled 'Governance in Action' of which she is the lead author (Nuijten *et al.*, 2004), an overview is given of the discussion about governance, after it became a popular concept in World Bank circles in order to re-evaluate the role of the state in development. This chapter is a very useful contribution to this debate. Not only does it provide an overview of the debate, it also makes a fundamental distinction between an instrumental and an analytical approach to the study of governance. Though others have also alluded to this distinction, this chapter draws the line between these two approaches in very concise and clear language. The instrumental is a goal-oriented approach, aiming at an ideal situation of good governance for collective goals and the deficiencies in current governance processes. The analytic approach to governance studies is defined by its focus on the actual process and not an ideal situation, aims at dissecting the divergent interests and consequences of governance processes for different groups in society (Nuijten *et al.*, 2004: 109-110).

The latter is precisely what Monique's work has always been about. Out of her engagement with those who experience the negative consequences of development, who are excluded from the benefits of economic growth, and who bang on the doors of the powerful, she points her analytical focus on the processes that produce these structural inequalities. Yet she does not go into abstract theorising for the sake of theory, she always keeps an eye on the human dimension, on the chaotic aspects of human life and the concerns of those who are at the receiving end. Her theoretical contributions are always linked to this human interest. This is also how I know her as a colleague: always interested in the human side of research, and in her colleagues and the well-being of our group. In short, she always puts effort in producing her own human force field around her from which we all benefit.

Han van Dijk / Sociology of Development and Change Group, Wageningen University, the Netherlands / han.vandijk@wur.nl

References

Nuijten, M., 1998. In the name of the land: organization, transnationalism, and the culture of the state in a Mexican Ejido. Wageningen University. PhD thesis.

Nuijten, M., 2003. Power, community and the state: the political anthropology of organisation in Mexico. London: Pluto.

Nuijten, M., 2004. Between fear and fantasy: governmentality and the working of power in Mexico. Critique of Anthropology 24(2): 209-230.

Nuijten, M., 2005. Power in practice: a force field approach to natural resource management. Journal of Transdisciplinary Environmental Studies 4(2): 1-14.

Nuijten, M., Anders, G., Van Gastel, J., Van der Haar, G., Van Nijnatten, C. and Warner, J., 2004. Governance in action: some theoretical and practical reflections on a key concept. In: Kalb, D., Pantsers, W. and Siebers, H. (eds) Globalization & development: themes and concepts in current research. Dordrecht Kluwer Academic Publishers, pp. 103-130.

15. THE LITTLE VOICE INSIDE YOUR HEAD: NUIJTEN, INTROSPECTION AND POLITICAL AGENCY

Bram Büscher

The enduring voice is often not the loudest. In a time when it seems that public debates consist increasingly of shouting matches between those who purposely do not want to understand each other[2], it is important to look for and listen to those voices that go against the grain, go unheard and actually have meaningful things to say. Monique Nuijten has been attentive to such voices and has herself been such a voice for a long time. In her work, she has been able to combine these gifts with sharp theoretical interventions to create an enduring legacy in political anthropology, urban studies and development sociology. This short essay will not be able to do justice to or give a comprehensive overview of this legacy. Rather, it aims to highlight what I believe are just some of the key theoretical, methodological and personal implications of Nuijten's attention to multitudes of different voices and how this helped her become such an important voice in her own right.

Monique Nuijten's interests in rural development are extremely broad. She has written on urban activism, legal pluralism, state bureaucracy, citizenship, corruption, food, imperialism, break-dancing and much more. Her work has been influenced by the

[2] Including on social media: the tweets of one 'PARTICULARLY OBNOXIOUS' world 'leader' come to mind immediately.

actor-oriented Wageningen development sociology, but never followed any 'party-line'. Nuijten is far too much of a free-thinker for that, especially given her consistent focus on being attentive to contradictions, tensions and using these to pursue novel theoretical avenues. Yet, throughout her different studies and field sites in Mexico, Brazil and Spain, I believe that two central pillars helped her navigate her diverse interests and develop several distinctive theoretical and methodological insights: introspection and political agency.

Starting with the latter, Nuijten's work centralises the agency of all subjects in her studies, but always in overlapping political contexts. With this I believe she means two things: first and foremost, that agents are always political subjects with interests, doubts, fears and desires but also, second, that agents always act within broader political contexts, from global hegemonic structures and development contexts to organisations and their bureaucracies. An illustrative intervention is Nuijten's astute article 'Between Fear and Fantasy: Governmentality and the Working of Power in Mexico', where she shows in ethnographic detail how both peasants and bureaucrats are part of, and indeed caught in, the interplay between the tension-ridden demands of formalised procedures and personalised relations. And it is precisely in this interplay that power expresses itself, according to Nuijten: it binds these actors together and leads them to develop strategies to cope with, change or resist this course of power. It is here that the relationality of power in Nuijten's thinking becomes especially apparent, something that she later conceptualised through the innovative term 'force fields' (Nuijten, 2005).

Trying to understand bounded dynamics and agential expressions in complex force fields has distinct methodological challenges, which Nuijten tackled through a combination of long-term ethnographic engagement and continuous inter-personal introspection. Following anthropological custom, for Nuijten there is no 'objective' positionality of the researcher. And it is here where the combination of political agency and introspection becomes particularly fascinating: as Nuijten builds up her familiarity of empirical force fields, she allows her own political agency to come to the fore in order to deepen her understanding. What I have in mind is how Nuijten describes what happened when she became more familiar with her interviewees and started to have more 'critical dialogues' with them, instead of formal interviews: 'I challenged people on certain ideas they held and deliberately confronted them with what I saw as contradictions in their statements and actions' (Nuijten, 2003: 23). She later reiterated this point methodologically: 'challenging people on certain ideas they hold and deliberately confront them with contradictions in their statements and actions can lead to important insights. It can be interesting to see how the research population reacts to the researcher's theories and doubts and to exchange personal views on the matter' (Nuijten, 2005: 11).

Importantly, what is 'interesting' here goes not only for the researcher, but also the research subject: critical reflection becomes iterative such that the researcher herself changes the force fields she tries to understand. Nuijten has consistently shown this to be a critical part of her research practices, most recently through her engagement with the Platform of Mortgage Victims (PAH) in Spain, whose often unheard voices she listened to, tried to understand and amplified through a booklet she shared generously with PAH members and allowed her to build enduring relationships. Hence, I would argue that Nuijten's research strategy is inherently personal as well: the way that Nuijten operates intellectually is to build up critical conversations and webs of meaning around things that truly matter *with* people rather than *around* people. One of the effects this strategy has for the people engaged in her research but also those reading her work, is that her theoretical and scholarly contributions combined with her personal approach have a tendency to become the proverbial 'little voice inside your head'; a voice that recurs, and helps to provide introspection, direction and understanding.

I have personally benefitted tremendously from Nuijten's insights, both academically and personally. Academically, I used her concept of 'force field' to understand both how power acts, but also how to approach power methodologically. Her concept was at the basis of the methodology for my Veni application and was instrumental in obtaining and implementing the project. Monique Nuijten was also one of the people who convinced me to apply for the professor and chair position of the Development and Change group at Wageningen University. I was hesitant at first, but Monique's belief in me and the chance to work more closely with her was one of the things that convinced me to go for it. Once in the position, I relied a lot on her guidance, collegiality and her reflections to navigate the 'force field' of the department within the broader university and to help pursue what was needed for the group and its individual staff members. Her active involvement in all facets of the group provided important bearings. In the office, too, she had a keen knack for unheard voices and things that needed to be heard by me or others. What often happened was that I was at work in my office and I heard a knock on the door. Monique would come in and ask to speak about something she noticed, heard or found important. This happened very regularly, and I remember those moment with great fondness, as Monique never came for trivial things: these were things she felt I needed to know. And nearly always, I came to agree fully.

I relied on and appreciated Monique's engagement and wisdom a lot when she was with us full-time but realised this even more when she suddenly had to stop working after the news of her brain cancer broke in the first week of 2018. I acutely felt Monique's absence; not only were we all shocked and worried about her, I immediately felt the absence of her intellectual and personal presence, and the way this helped me – and I suspect most of us in the group – to see things more clearly, to have to face my contradictions and be attentive to the voices that were not sufficiently heard. I still sometimes sit in the office thinking what Monique would say, or what voices she believes I should pay more attention to.

Referring to Monique Nuijten's contributions as 'a little voice in your head' perhaps sounds slightly belittling to some, as though Nuijten's voice was 'little'. But I want to argue it means exactly the opposite. One of the things Monique taught us is that it is the marginal, 'little' or unheard voices that you need to pay attention to, that you must take seriously. And I think she is right: you often forget the loud, annoying voices that have little meaningful to say, but you remember those voices that become part of your unconscious; that speak to you even when you're not necessarily paying attention to them; that become truly meaningful in many different ways. Monique Nuijten is such a voice, one that will endure for a very long time.

Bram Büscher / Professor and Chair, Sociology of Development and Change Group, Wageningen University, the Netherlands / bram.buscher@wur.nl

References

Nuijten, M., 2003. Power, community and the state: the political anthropology of organisation in Mexico. London: Pluto.

Nuijten, M., 2004. Between fear and fantasy: governmentality and the working of power in Mexico. Critique of Anthropology 24(2): 209-230.

Nuijten, M., 2005. Power in practice: a force field approach to natural resource management. Journal of Transdisciplinary Environmental Studies 4(2): 3-14.

16. SUFFERING FROM FORCE FIELDS

Michiel Köhne

The concept of force fields connects resources to the ways in which their use is negotiated and to what this means to people, by foregrounding how more structural forms of power shape local people's lives (Nuijten, 2005). Monique Nuijten's concept has inspired many, including myself, to study local resistance against resource exploitation. In Australia, in the small town of Comley,[3] a battle raged over coal mining between a forceful group resisting the mine, mostly anticipating harm to the environment and their health, and an equally forceful group supporting it, mostly hoping for jobs and coal money coming to town. As a tribute to Monique's work on resistance I here depict the different meanings coal acquired for opponents of the mine as a result of the fight over it. I do so by focussing on how these meanings are mediated by the local force fields constituted by the requirements of social expectations of family, friendship and community.

[3] All people and places have been anonymized to protect research participants.

The beginning

The mining industry entered Comley covertly, silently buying up land one farm after the other, playing neighbours against each other by negotiating under strict secrecy. Farmers were seduced by high prices while being frightened of a future in which nobody else would be interested in buying their land. A future in which farmers would have left, leaving an open pit coal mine and a mining town going from boom to bust. For these reasons, 'No-mine Comley', a group of inhabitants opposing the mine, tried to get farmers to commit to refuse to sell collectively. Angus is one of the farmers who was torn between the forcefield constituted by the coal mining buying power and the local forcefield of farmers solidarity. However, when people around his property began selling their land, he lost his enthusiasm for his own farm and sold up. It made him really sad, 'it takes away your family history, money is not an answer to all this [...] the effects on people may be worse than the effects on the environment, it's tearing us apart.'

Dealing with the divide

Most people tried to avoid being confronted with the divide in the community. Many opponents therefore evaded the pro-mining people. Kylie, who lives on a farm on the perimeter of the mining area, is one of them: 'This certainly has stopped us going into town [...] some shops were hoping to make fortunes from the coal mine workers.' Henry, her husband: 'I still rarely go to the cattle sales, cattle men just think of jobs and I feel resentful about some not speaking up for me, it's hard to tell what they think of me.'

However, at some places local force fields mediated this evasion by imposing its own social rules. For example, churches and sport clubs were considered neutral territory. Jack, a priest, explained how mining supporters and opponents often sat side by side in church, deliberately not talking about it. 'It hurts, not to be able to talk about what's keeping you busy most of all, it really detracts from the feeling of community in the church, yet it's better than openly arguing.' Ned and Victoria explained to me how mining had become the elephant in the room: 'that is the easiest, just not talk about it, [...] but I do feel a loss of connection.' Others just couldn't stand the tension of not talking about it and stopped going to church. In sports clubs, too, mining was strictly avoided as a topic of discussion. It was even more difficult there, however, because many had accepted financial support by the mining company. Still, the sense of community enforced by the force fields of both church and sport clubs meant not talking about it.

Closer to home, communal interests sometimes superseded the fight over the mine. Stephen, who has been a zealous protestor for 15 years, lives on a fairly remote farm with immediate neighbours being staunch supporters of the mine. They depended on each other for many things, so they avoided talking about mining: doing maintenance work to their common fences, participating in the local fire brigade and even celebrating Christmas together. 'You just need a reasonable relationship with your neighbours,' he explained, 'We get our milk from him and he grazes his cattle on our land. We have a common Christmas Party and enjoy this as we've known each other for years already, but no one mentions the word.' Local force fields of community and good neighborship thus affected the meaning of the divide caused by the coal.

Consequences of the divide for resistance

Local force fields of family and (religious) community also affected how opponents resisted the mine. Jack is dead against the mine but could not mention this in his church: 'We [...] had conversations with activists, about why we couldn't be vocal about it or join the protest, but it's a small town, you gotta be nice to people. I felt a bit of a hypocrite, I loved that others did it.' Sheila also refused to become more activist: 'I don't want to be seen as activist [...] I don't want to run the risk of being ostracised.'

Many people struggled to balance their environmental conscience with the force field constituted by the daily requirements of family or community life. Shane's daughter had a health condition that did not allow them to live next to the mine, but their property value had gone down, so they could not afford to sell and leave either. They deliberately restricted their activism as their only hope if mining were to proceed was a buyout by the company. Also, the atmosphere in the house had suffered from their frustrations about losing the battle and their kids socialised with pro-miners' kids. That's why they felt they had to adopt a more neutral position, 'so they cannot attack us'. Poppy and Banjo also tried to maintain a neutral position: 'When there is a market in town, they push it in your face, 'sign our petition!' We now sign both pro and against mining petitions, so we can't upset anybody.'

Others who did protest, sometimes suffered from pressure of the force field of community or family expectations. Alice told me how she felt, protesting in the main street: 'I walked the march but hid myself with a cap and sunglasses. They seem to hate protestors. I have never told my parents [...] In the march, I felt sick in my stomach. I didn't know what repercussions it would have. [...] I did it just once, I felt so sick [...] Now, I feel ashamed of myself, why haven't I been more active?' Often male family heads called upon this shame to stop their daughters or wives from activism. As Mia related about a close friend, 'She came to me crying, she wanted to walk the protest marches with us, but the men in the family put heavy pressure on her, her father said that she would be embarrassing the family.'

Many of those who chose to be vocal have borne the brunt of the forcefield constituted by friendship and family. Mia explained how the mining conflict entered her family: 'My husband likes to go have a beer with the boys, but he doesn't go anymore since I was protesting. At first, he made a joke of it, telling me about the remarks that he got 'can't you control that wife of yours?', but later he came home asking 'can't you stop it?' Karin also lost some of her old friends: 'My main friends were people who were born here, some of them I lost. They would swear at me on the road 'you are f-ing up Comley! We need jobs!'

Personal consequences of the divide

However, people's reaction to the divide over coal was not only mediated through local force fields. It also constituted its own force field, affecting protestors personal life, for example how people experienced their own community. Sheila explained: 'Because of the divide anybody who isn't us, is them. It really reduces community spirit, it's always there: who is pro and who is against.' And Alfred, one of the activists at the forefront commented on how the community had changed: 'We're coping by not interacting.' Dominique told me how she felt she could not trust anybody anymore. 'I can understand how people become fundamentalist. [...] After some time, the only people I felt comfortable to be with were the people that joined the monthly marches downtown.'

The divide was most severely suffered by Matilda, the leader of No-mine Comley. For hours, she told me stories of what it meant to be at the forefront of the fight. Her being bullied started at the school playground while waiting for her children and gradually extended to the point where even people she had always felt comfortable around started avoiding her, saying: 'I cannot support No-mine Comley, because I will be ostracised, but I can also no longer be seen with you.' Coal has completely changed her life, losing some friends while making others, but mostly by being hated by many of the supporters of the mine. As a leader of the protests she bore the brunt of the force field emerging from the divide.

The force field constituted by the divide has also changed people's view on life. Many activists became resentful. As Kylie said: 'It makes you distrust everybody. I have become bitter and twisted.' Various people also explained how their lives had been put on hold, or 'in maintenance mode', as Henry liked to call it: 'I'm still fixing up the fences, but you're in limbo, we really can't think about anything until we know what is going to happen.'

For quite a few opponents, especially the more activist among them, the stress endured under the force field of the divide made them feel like giving up: just move away and start somewhere else anew. Jenny told me why she moved 40 minutes away, 'the idea was simple, no mine, no fight.' She felt her life was completely determined by the divide, 'the invasion of it, not just the physical, but the invasion of minds and relationships'. Others have stayed, like Ned: 'At times we have considered leaving because [...] of how uncomfortable our social life has become, at times you just live and breathe the conflict [...] When you woke up, the first thing you think is mining.' Leaving was not an option for everyone. Poppy explained, 'If the coal goes ahead most of my friends would leave, but I would not be able to afford to leave.'

Concluding

Local force fields of family, friendship and community considerably affected the meaning of coal mining for the people of Comley. However, the coal mine also lead to a split of the community between proponents and opponents that in itself constituted a force field affecting the meaning of coal mining as it changed people's life on a personal level. Monique's notion of force fields in this way helps to understand how coal affected its opponents not only by force fields regarding their position towards the mining company but also by force fields around local relationships.

Monique is valuable to academic life because of her conceptual contributions on force fields and others, but also because of her person. A few years ago, I asked Monique Nuijten for advice on new directions for my research. After listening to my ramblings on what I didn't know, but what I thought was important, and on how I didn't know where to start, she gave me a very simple answer, 'me too, I don't know either, just keep going.' Thank you, Monique, for helping me this way.

Michiel Köhne / Sociology of Development and Change Group, Wageningen University, the Netherlands / michiel.kohne@wur.nl

References

Nuijten, M.C.M., 2005. Power in practice: a force field approach to power in natural resource management. Journal of Transdisciplinary Environmental Studies 4(2): 3-14.

17. 'URBAN NUIJTEN': THEORY, ETHICS, AND COMMITMENT IN THE ACADEMY[4]

Arturo Escobar

Working 'from below' means being attentive to modes of interpretation and explanation that do not necessarily privilege the discourses used in the public media, ones articulated by governing agencies or for that matter those shaped by theoretical debates.
Nuijten, 2015: 481

In these brief notes, I will focus on Monique Nuijten's writings on urban themes what we could call 'urban Nuijten' –, which I discovered only recently, and which unveiled for me a whole series of interests, conceptual interventions, researches, and life politics on Monique's part of which I was not aware. Despite continuities (largely stemming from her commitments to political anthropology and her deep concern with human suffering, subalternity, and resistance), I think that, taken as whole, the set of field research projects and papers dealing with urban issues constituted a significant

[4] My thanks to Elisabet Rasch and Pieter de Vries for sending me the papers by Monique on which this text is based, and to Elisabet and hear colleagues at Wageningen University for inviting me to reflect on the meaning and importance on Monique's work in celebration of her life.

departure – perhaps even a sort of reinvention – on Monique's part as a scholar and as anthropologist, in relation to her substantial previous work in Mexico. I am not suggesting that this facet of her work is homogeneous; in fact, the opposite is true, which makes it more exciting (and which, I imagine, made it more exciting for her and her collaborators).

At the basis of this body of work, I think, there is an urban-political and existential orientation, which comes through clearly in one of Monique's most recent articles on a group of break dancers from a poor neighbourhood in Quito (Bode Bakker and Nuijten, 2018). In this case, the notion of 'utopian aspirations' becomes a medium to explore the avatars of young people's subjectivities, on the one hand, but also how for the dancers art becomes a medium for the celebration of life, in their steadfast refusal to conform to the restrictive scripts society has in store for them, which assume little education and employment, if at all. At issue is the question of how, in becoming break dancers ('b-boys and b-girls'), these youth craft a meaningful existence in contexts in which their lives have little or no value. By constructing a disciplined and highly practiced body, the dancers turn their bodies into a 'revolutionary text' which contributes to create a positive imaginary of life on the streets and a potential alternative to the much-touted middle class life (a very socially and ecologically damaging model in Latin America at present, I must add, centred on individualism, expansive territoriality, and consumption). The body is central to this ethics of street art, but not only that, for along with the body the soul dances, too, as a way to ease their emotional pain, as beautifully expressed by one of the dancers ('[When breaking] you make your soul [*alma*] dance,' p. 221). For the few young women who venture into the break dancing terrain, becoming a 'b-girl' involves the creation of new images of femininity and of independent selves, in their own unique way. For them, break dancing becomes a source of empowerment and transformation in relation to established gender subjectivities, the space for a feminist feminine practice of their own, one that challenges the heteropatriarchal norms they find so stifling.

This seemingly stand-alone paper, however, has theoretical affinity with the rest of Monique's urban work. One may say that what brings her various urban projects together is her concern with the lives and agency of poor people in popular neighbourhoods and, on the theoretical side, her resolve to find enabling ways to understand their agency as they strive to deal effectively with the multiple situations and forms of subordination shaping their lives. In this way, working in an altogether different geo-political and onto-epistemic space and research field – a poor neighbourhood in Recife, Brazil, undergoing redevelopment –, allows her to propose a cogent explanation for why poor people in the urban Global South are so suspicious of any attempt at 'inclusion' and 'participation,' whether by the State or by development NGOs, whether from the right or from the Left, while developing meaningful urban practices of their own. This suspicion is born from their engagement with projects that often times leave them newly marginalised or damaged along the way. Her ethnography focuses on the process by which the economically and ecologically precarious lives of the urban poor become the stuff of modernist resettlement schemes by the World Bank and regional and local governments intended for the spatial 'modernisation' and 'civilising' of informal settlements, in ways in which, despite the rhetoric to the contrary, the poor end up being sidelined in project design and implementation. Neoliberal notions of 'citizenship' and 'security' (always racially coded in Latin America) provide the discursive space for such slum upgrading projects even in the allegedly Left context of Worker's Party governments, resulting in 'citizenship games' that she has no qualms to characterise as 'perverse' (Nuijten, 2013: 22).

Rather than seeking to expand the notion of citizenship, as most other scholars working on similar contexts have done, this realisation leads Monique to take a more radical position, that of '[going] beyond the citizenship framework if we want to give serious attention to the perspective of the poor' (Nuijten, 2013: 10). (Incidentally, this reminded me of what a brilliant black activist from the Colombian Pacific once told me, commenting on the rhetoric of citizenship and inclusion: that 'to become citizens would be akin to returning to the time of slavery.') The political anthropology emphases on poor people's political subjectivities and the relations to the State thus become, in Monique's radicalised framing, an opportunity to explore citizenship as a particular language of the political, widening the space of the political itself to include a whole ensemble of popular idioms. This could only come about because of her ability to always 'think from below,' and from there query our most cherished scholarly perspectives.

The theoretical underpinnings and relevance of this work had already been explored in a joint paper with Martijn Koster and Pieter de Vries based on the same Recife project. Here, the notions of spatial ordering, neoliberal regime of citizenship, and the analysis of peripheral modernities provide the basis for examining the 'entangling of rationalities' (Nuijten *et al.*, 2012: 166) at work in the making of urban space. The view of the project that emerges is indubitably clear: 'One could argue that such slum upgrading projects, instead of looking at what the poor need, demand that they integrate into mainstream society and follow middle class norms of how to behave and use space – but without the jobs, money and other attributes for maintaining such a lifestyle' (Ibid.: 166). This realisation, nevertheless, would be incomplete if not coupled with an incisive analysis showing how people make do with the discursive and material infrastructures laid out by the slum upgrading project itself. While building on earlier works highlighting subaltern forms of agency (e.g. James C. Scott, M. de Certeau), these studies anticipate the important 'relational turn' in recent urban studies (Amin and Thrift, 2017; Simone, 2019; Simone and Pieterse, 2017).

Fast-forward to Monique's most recent urban work, this time in Madrid. The new field site allows her to explore in new ways her long-standing political anthropology interest in grassroots political agency. 'The assumption in my work,' she states at the outset, 'is that people always actively challenge and reflect upon the political conditions of their lives.' It is important to underscore that this seemingly indisputable anthropological statement is actually inimical to most social movements' studies in political science and sociology, the disciplines that dominate the field. In these approaches, as is well known, subaltern knowledge is sidelined and, often times, completely ignored in the scholarly accounts. Simply put, common people, and their activists and organisations, are not seen as producing knowledge relevant to the scholarly concerns. And it is to activists as knowledge producers that Monique turns in this project, without overlooking her long-standing interest in people at the margins.

Taking as a point of departure Franz von Benda-Beckmann's proposition that engaging with the law as a situated, lived-in apparatus requires long-term engagement, Monique's project with Madrid's Platform of Mortgage Victims (PAH), which she started in 2014, was theoretically very much about power and agency, whether through von Benda-Beckmann's important work on multiple co-existing normative orders, including but well beyond modern law (which, I must confess, I did not know about), Eric Wolf' injunction to study 'the flow of action' that accompany power dynamics, or Norman Long's notion of interfaces as domains where power finds particularly acute expression. Her poststructuralist conceptualisation of the political (influenced by Chantal Mouffe's notion of antagonisms as the pivot of social life), leads her to see power in terms of the encounter of normativities in situations of conflict. It couldn't be otherwise when dealing with grassroots politics; as her analysis so persuasively demonstrates, activists mobilise in fields of clashing normativities, hoping to shift values in the direction of greater justice. The PAH, started by Ada Colau in Barcelona in 2009 as an organising strategy to counter the wave of house evictions in the wake of the collapse of the housing bubble in Spain, presented for Monique a fertile ground on which to explore these issues. Above all, it seems to me, Monique elegantly demonstrates why grassroots politics is a rich terrain in which to investigate the constitution of the political, taking neither State law nor pre-given modernist notions of the political as the unexamined onto-epistemic background for the investigation.

The 'urban Nuijten' contains an incredibly rich set of inquiries for thinking about a whole set of increasingly salient urban problems, from poor people's encounter with the police (central to the Black Lives Matter movements in the US, Brazil, and Colombia, among other countries) to questions of subaltern legalities, urban design, poor people's relation to the State, and urban social movements. Her work in Brazil with Pieter de Vries and former students has essential lessons for understanding the onto-epistemic structures underlying the devaluation of Black lives in Latin America and the profound anti-Black racism in the Continent. It also shows why urban improvement schemes make more sense for local bosses and politicians than for the poor themselves, whose lives often really don't matter all that much. Above all, it illustrates why these schemes are ultimately about the continued appeal of modernity as an inexorable attractor of people's desires, but also people's ambivalence towards it, particularly because of what such 'modernity' comes to be in real life through strategies such as slum upgrading schemes or many other allegedly poor-focused development projects. Finally, urban work centred on grassroots agency today might be really useful in coming up with strategies of resistance to the implicit imposition of middle class models of habitability as default setting in urban design and modernisation schemes, while revealing the limits of inclusion and participation for the poor under any urban/spatial-political regime existing at present.

It seems to me that Monique's engagement with Latin America doesn't stem from any romantic attachment or political utopia, but is deeply mediated by an acute consciousness and concern for the lives of the 'wretched of the earth' in the countries and communities where she worked. An essential part of her scholarly work is the practice of what I would call a re/constructive approach to theory (as in the case of her treatment of von Benda-Beckmann in her PAH article, Nuijten, 2015). Her concept of 'languages of the political' is a significant reconstructive effort at re-theorising the political, and a unique contribution to anthropology and beyond.

Engaging with 'an entanglement of rationalities' is perhaps what defines the project and promise of anthropology as a whole. Critical of any and all forms of rationality that seek to impose themselves on other ways of worlding, we hope to make available to others, particularly to those under the weight of power, the best maps we can draw of the complex interplay of worlds and their attendant rationalities, so as to make the world a better place for all, as we ourselves strive to live ethically and politically in between worlds. It seems to me this has been a crucial dimension in Monique's own brand of 'working from below,' as the epigraph with which I started this text suggests, and this is a lesson, I believe, we all can take to heart in contemporary worlds so enamoured with power, including the often pyrrhic forms of power we, academics, seem to find in what we do. For thinking from below is possibly the best antidote against such attachment to power in our trade.

> *In a time of turmoil and crisis for a large part of the world population, and increasing environmental concern, a variety of initiatives have emerged at the grassroots, based on values which differ from those of neo-liberal capitalism.*
> Nuijten, 2015: 477

Arturo Escobar / Professor of Anthropology Emeritus, University of North Carolina, Chapel Hill, USA / aescobar@email.unc.edu

References

Amin, A. and Thrift, N., 2017. Seeing like a city. Polity Press, Cambridge, UK.

Bode Bakker, M. and Nuijten, M., 2018. 'When breaking you make your soul dance' Utopian aspirations and subjective transformation in breakdance. Identities 25(2): 210-227.

Nuijten, M.C.M., 2005. Power in practice: a force field approach to power in natural resource management. Journal of Transdisciplinary Environmental Studies 4(2): 3-14.

Nuijten, M., 2013. The perversity of the 'citizenship game': slum-upgrading in the urban periphery of Recife, Brazil. Critique of Anthropology 33(1): 8-25.

Nuijten, M., 2015. Political activism and the PAH (platform of mortgage victims) in Spain: the significance of Franz von Benda-Beckmann's work for the study of power, political agency and legal pluralism at the grassroots. The Journal of Legal Pluralism and Unofficial Law 47(3): 476-492.

Nuijten, M., Koster, M. and De Vries, P., 2012. Regimes of spatial ordering in Brazil: neoliberalism, leftist populism and modernist aesthetics in slum upgrading in Recife. Singapore Journal of Tropical Geography 33(2): 157-170.

Simone, A.M., 2019. Improvised lives. Cambridge: Polity Press.

Simone, A.M. and Pieterse, E., 2017. New urban worlds. Inhabiting dissonant times. Polity Press, Cambridge, UK.

18. OLD AND NEW URBAN QUESTIONS

a personal response to Nuijten *et al.* (2012), 'Regimes of spatial ordering in Brazil: Neoliberalism, leftist populism and modernist aesthetics in slum upgrading in Recife.'

Robert Coates

Monique Nuijten's lead-authored 2012 article on displacement of the urban poor in Recife, Brazil is an important contribution to the urban development literature, as well as to debates on the subjectivities, strategies and tactics generated under 'high modernist' urban policy and aesthetics. By intersecting a concern for neoliberalism's 'progressive' extension under Brazil's Workers' Party (2003-2016), with ideals of modernity that have long framed 'slum upgrading' policies *and* the social dynamics generated in their wake, Nuijten *et al.* provide a ground-breaking exploration of mainstream development thinking in the early 21st century. The article places the authors' fingers on the pulse of critical currents in contemporary development praxis.

In this response to the article, I draw out a few themes that resonated with me personally, as well as ask *what we know now* that could take these ideas forward. I focus specifically on two areas. Firstly, a concern with environmental crisis that I believe leads us to reframe early 21st century literature on 'the agency of the poor' – and that has, in the last decade, influenced more and more the justification for urban removals in Brazil and beyond. I provide here added interest in urban political ecology *and* the temporal politics of 'future risk', that influences considerations of urban space-making and subaltern subjectivities. Second, I suggest how Nuijten *et al.*, in some ways, foretell the changes and dissatisfactions underpinning progressive neoliberalism in Brazil, that provided such fertile ground for far-right nationalist manipulation amidst the urban working class in the country's new authoritarian populist phase.

Nuijten *et al.* zoom in on Prometrópole, a World Bank-funded slum-upgrading scheme begun in Recife in 2003. Typical of high modernist infrastructural works designed to instil 'Order and Progress' (as the Brazilian national motto goes), the project aimed at ridding the city of *palafitas* – poor, favela-type housing next to city waterways – through capital-intensive river canalisation and drainage and the rehousing of residents in identikit blocks. The authors' extensive ethnographic work undertaken over a number of years points us toward the discourses underpinning project implementation, as well as to residents' reactions. Drawing on Scott (1998) they show that the scheme was based on the 'aspiration to improve the human condition through 'a sweeping, rational engineering of all aspects of social life' (p.158). Simultaneously, using Berman (1982), they argue that the powerful modernist imagery associated with 'upgrading' appealed to slum dwellers' own desires for life improvement, while also, in a Foucaultian sense, coopting them into hegemonic strategies that were bound to fail in their stated objectives, yet of course remain deeply successful for extending neoliberal state power.

While the interest in power and knowledge abounds, the article also draws on De Certeau's (1984) theory of everyday life to suggest how resettled populations tactically work to 'face and contest, or at least evade, the disciplinary forces of the strategies inherent in regimes of urban renewal' (p.167). Project beneficiaries avoided meetings with organisers and funders, and ignored the constrictive rules of their new highly-ordered blocks by personalising their apartments' design, appropriating space outside them, or even by selling off their contents and renting them out while undertaking new land occupations elsewhere. 'They just don't obey', exclaimed an exasperated official (p.165), flummoxed that order and progress remained as out of reach as ever – yet (I would strongly wager) reinforcing their belief that the poor masses were beyond help. Foucaultian scholars would of course view these resistances as part and parcel of the circulatory workings of political power itself. Monique and colleagues, nonetheless, interpret this as something more agential, demonstrating residents' disruption of modernist narratives that *they know* '[gloss] over…the real causes of poverty and exclusion' (p.165).

Interest in the lived experiences, 'tactics' and 'life world' (p.167) of the poor no doubt reveals something of the authors' earlier training in Wageningen actor-oriented sociology, but now demonstrating a clearer focus on the discursive forces underpinning the production of deeply uneven urban space. In this sense the article aligns with important geographical and anthropological writing on what Ballard (2015) calls 'development *by* the poor' (as opposed to development *for* or *without* the poor) – which in the last two decades spawned such phrases as 'insurgent citizenship', 'the quiet encroachment of the ordinary', and recognition of 'the right to have rights' (c.f. Bayat, 2000; Caldeira, 2017; Dagnino, 2005; Holston, 2008; Lund, 2016; Scott, 1998). The authors show that we ignore poor peoples' perspectives and actions, as they impact upon the state itself, at our peril. Yet, surely, we must also consider within these dynamics the material production of (planetary) space in seemingly endless concretised densification, with deep inequality accompanying lifeless waterways-turned-sewers, ever more frequently flooding the inhabitants at their margins. As Brazilian geographer Mauricio Abreu (1994) argued, 'the dance of the favelas' is *itself* constitutive of the changing material environments of urbanisation. The displaced poor routinely occupy further land in the process of appropriating and subverting modernist interventions, with the state 'returning' to displace them in perpetuity. The urban poor's 'rhythms of endurance' (Simone, 2018) might be seen to reside on the same shaky ground as the modernist state itself (Coates and Garmany, 2017).

But there is more to the story. Monique and colleagues also provide a timely intervention into what Peck and Tickell (2002) termed 'rollout' neoliberalism: state-led *re*-regulation geared to extending free markets, and a base citizenship defined by individual economic rationality and personal responsibility. After the rollbacks of the 1980s and '90s, this was typified by the rise of left-liberal governments like that of Brazil's Workers' Party (PT), who walked a tightrope between on one hand appeasing market growth and expanding the wealth of Brazil's super-elite, while on the other introducing wide-ranging social welfare and cash transfers. Again drawing on Scott (1998), the authors show how it was *progressives* that introduced Prometropóle, with Recife's PT government motivated by the archetypal promise of large-scale infrastructure '[transforming] people's habits, work, living patterns and moral conduct' (Scott, 1998: 89, cited Nuijten *et al.*, 2012: 158).

A decade later and faced with the emphatic social and environmental disaster that is the Bolsonaro government, we must surely be cautious of over-criticising the PT. Yet probing the 'progressive neoliberal' era rather than the party politics represented by it should also be fruitful for asking *how could it possibly have come to this?* Nuijten *et al.* (2012) give us some clues over how the electoral constituency that Bolsonaro built in 2018 became possible. We know that Bolsonaro's base support was the (upper) middle class – the 'traditional' right wing was decimated as supporters pushed further right – and yet the working class votes he took from the PT were essential to his election. While this was especially prevalent in Brazil's south/southeast and in interior states, and notwithstanding Recife and the northeast maintaining PT majorities, there was still a significant swing against the PT right across their urban heartland.

Critical analysts point (especially) to Bolsonaro's support for evangelical churches and his supposed firm hand against political corruption and violent crime as key factors (Richmond, 2020; Saad-Filho and Boffo, 2020). However, as McCarthy (2019) and others have discussed, it is hard not to view this in the shadow of the failures of progressive neoliberalism that left poorer urban classes – especially after 2010 as the 2008 financial crisis bit hard – facing 'personal responsibility' for declining job prospects and share of national product, deteriorating environmental quality, and higher crime. In short, a decade of the PT had not brought forth the modernity all had hoped for, and disappointment left poorer classes splintered, demotivated, and susceptible to a formidable propaganda machine on social media. In hindsight, Nuijten *et al.*'s (2012) article was documenting progressive neoliberalism's endgame, just as a new dynamic of free-market populist authoritarianism waited in the wings.

Relatedly, and as a final substantive point, I want to address the changing nature of urban displacement in which, increasingly as the years advance, people are evicted due to climate or environmental risk. According to the Forced Migration Observatory (https://migracoes.igarape.org.br/), the rationale for the vast majority of 7.7 million Brazilians forced from their homes since the year 2000 has been 'natural disaster risk'. Historians of Brazil might want to deny this represents any major change, and point to longstanding 'hygienist' narratives in which the urban poor were *always* viewed as the locus of risk, disease, and environmental ill-health. Prometropóle-type interventions were given impetus by the teleology of 'rising out' of a barbarian past (e.g. Garmany and Richmond, 2020). Yet the sheer scale of recent evictions – and of the disasters they usually follow – begs a nuanced understanding of the political ecologies of urban security that underpin such processes.

The key here is to understand changing narratives of urban time *as well as* of urban space. As a number of authors argue, displacements increasingly focus not on bringing favelas *out* of pre-modernity but rather on avoiding an insecure *future* characterised by 'emergent' threats like climate change (c.f. Anderson, 2012; Zebrowski, 2019). Pre-empting disaster *after* (flood and landslide) disasters have actually struck of course represents an extraordinary turn in urban political discourses, consummate to our extraordinary political and environmental times. Zebrowski (2019) equates this change to 'event suppression', precisely because turning quickly to displacement prevents a public deconstruction of *why* the flood or disaster occurred in the first place. Discourses of 'restoring normality' or 'building resilience' remove the 'venom' (ibid.: 158) that disaster shocks present to political order, and ensure a future of enhanced business-as-usual. Again with hindsight, and though I know of Prometropóle only what Monique and colleagues outline, could it be that the scheme also carried the utterings of a change in urban political temporalities, beyond spatiality alone?

Given the still apparently unfettered belief in urban economic growth that characterises mainstream politics, we are right to question the motivations and discourses behind well-funded projects that displace poor settlements and seek to rehouse them in 'hygienic' symmetrical blocks that are now common all over Brazil. Nuijten *et al.* (2012) provide a vivid description here of the responses and resistances of resettled populations, who clearly aspire to improvement yet never simplistically reproduce the behaviours demanded of them by neoliberal 'high modernism'. A decade on, and faced with the twinned dynamics of authoritarian populism and embedded urban socio-environmental risk, exploring the links between material environments and political subjectivities remains of the utmost importance.

Robert Coates / Assistant Professor, Sociology of Development and Change Group, Wageningen University, the Netherlands / robert.coates@wur.nl

References

Abreu, M., 1994. Reconstruindo uma história esquecida: origem e expansão inicial das favelas do Rio. Espaço e Debates 14(37): 34-46.

Anderson, B., 2010. Preemption, precaution, preparedness: Anticipatory action and future geographies. Progress in Human Geography 34(6): 777-798.

Ballard, R., 2015. Geographies of development III: militancy, insurgency, encroachment and development by the poor. Progress in Human Geography 39(2): 214-224.

Bayat, A., 2000. From 'dangerous classes' to 'quiet rebels': the politics of the urban subaltern in the global south'. International Sociology 15(3): 533-557.

Berman, M., 1982. All that is solid melts into air: the experience of modernity. Penguin Books: New York.

Caldeira, T.P., 2017. Peripheral urbanization: autoconstruction, transversal logics, and politics in cities of the global south. Environment and Planning D: Society and Space 35(1): 3-20.

Coates, R. and Garmany, J., 2017. The ecology of citizenship: understanding vulnerability in urban Brazil. International Development Planning Review 39(1): 37-56.

Dagnino, E., 2005. Meanings of citizenship in Latin America. Brighton: IDS.

De Certeau, M., 1984. The practice of everyday life. University of California Press: Berkeley.

Garmany, J. and Richmond, M.A., 2020. Hygienisation, gentrification, and urban displacement in Brazil. Antipode 52(1): 124-144.

Holston, J., 2008. Insurgent citizenship: disjunctions of democracy and modernity in Brazil. Princeton: Princeton University Press.

Lund, C., 2016. Rule and rupture: state formation through the production of property and citizenship. Development and Change 47(6): 1199-1228.

McCarthy, J., 2019. Authoritarianism, populism, and the environment: comparative experiences, insights, and perspectives. Annals of the American Association of Geographers 109(2): 301-313.

Nuijten, M., Koster, M. and De Vries, P., 2012. Regimes of spatial ordering in Brazil: neoliberalism, leftist populism and modernist aesthetics in slum upgrading in Recife, Singapore Journal of Tropical Geography 33: 157-170.

Richmond, M.A., 2020. Narratives of crisis in the periphery of São Paulo: place and political articulation during Brazil's rightward turn. Journal of Latin American Studies 52(2): 241-267.

Saad-Filho, A. and Boffo, M., 2020. The corruption of democracy: corruption scandals, class alliances, and political authoritarianism in Brazil. Geoforum: in press.

Scott, J.C., 1998. Seeing like a state: how certain schemes to improve the human condition have failed. New Haven: Yale University Press.

Simone, A., 2018. Improvised lives: rhythms of endurance in an urban south. John Wiley & Sons.

Zebrowski, C., 2019. Emergent emergency response: speed, event suppression and the chronopolitics of resilience. Security Dialogue 50(2): 148-164.

PART 2:

COLLEGIAL ENCOUNTERS

In which eleven colleagues describe the personal qualities that define Monique Nuijten as colleague, mentor or supervisor, drawing a portrait of Monique as a unique scholar who redefined teaching, supervision, mentoring and collegiality as (more than) academic pursuits. An image emerges of Monique Nuijten as a humble yet strong-willed scholar able to transcend and connect the personal and the professional, the theoretical and the empirical.

19. INTERVIEW WITH MONIQUE IN THE ACTOR-ORIENTED EXPRESS[5]

Gerard Verschoor

Me: 'Dreary day out there – so cold and wet! Nice and cosy in here. Feels warm. Glad we took this train!'

Monique *[laughs]*: 'A good setting for an interview, yes.'

Me: 'Yes… This interview is about your life and work, but we have a few stops only before our final destination. Two thousand words, max. Can we do it?'

Monique: 'Too short for all of that. Maybe about my work only?'

Me *[taking a deep breath]*: 'You mean we skip forty years of personal anecdotes and layers upon layers of meaning and experience? Like when Margreet was asked to play the Virgin Mary in La…'

Monique *[laughing, interrupting]*: 'La Canoa. Yes… Fond memories. A lifetime full of them!'

[5] A totally fictitious, could-have-been interview that never took place.

Me: ' Yeah… Maybe we shouldn't do this. I think Emily Dickinson once said that saying nothing sometimes says the most.'

Monique: 'No. It's fine. Let's talk shop.'

Me: 'As we always did! Perhaps a bit of politics, too?'

Monique *[resolutely]*: 'No! Not this time'.

[Screeching noise of wheels on rails as train breaks for a first stop]

Me: 'Fine. We concentrate on the work then. Seems we can't do without it. Come to think of it: in all these years we *didn't* talk shop maybe a handful of times only…'

Monique: 'There was, there is, so much to say!'

Me: 'Agree. So let's get on with it. Let me go straight at it. We all hopped on this actor-oriented thing Norman Long started moulding in Wageningen. We went to Mexico to carve things out further: Hans, Elsa, Jikke, Humberto, Dorien, Alberto, Magda, Pieter, Horacia, Gabriel, our silverback…'[6]

Monique: 'Yes. We went there with a mission: not to teach the people we engaged with how to interpret why they were doing what they were doing, but to learn from them how we should talk about them in the first place…'

[Whistle blows; dry thump as doors latch]

Me: 'I guess this was the central issue. Come to think of it: in all this collective work nobody came up with an externalist explanation of why people were doing what they were doing in the way they were doing it. Which sounds like a pretty Mancunian idea to me. Nevertheless, an outrageous proposition at the time…'

Monique: 'Right. I think it even was kind of revolutionary: to acknowledge that farmers, *canaleros*, or the rank-and-file from the line agencies had problems that were radically different from our own, and that it was impossible to predefine what these problems were from theoretical mandates keen to obtain Grand Vistas through unifying explanatory schemes. And saying so didn't make us much friends. Not in the beginning, at least. At the time, the sociology of development was full of self-aggrandising, self-indulgent types who thought that it was a great virtue to come up with explanations that came straight from another time and context, and not equipped really to deal with the more

[6] Hans Heijdra, Jikke Verhulst, Humberto González, Dorien Brunt, Alberto Arce, Magda Villarreal, Pieter van der Zaag, and Gabriel Torres participated in a project to develop the actor-oriented approach on planned intervention between 1986 and 1992.

globalised setting we were experiencing. Yet most of them had never set foot in a rural community for more than a weekend!'

Me: 'Well, there were people like Luis[7] at Colmich[8] who weren't that pretentious and *did* do the fieldwork.'

Monique: 'Sure. And they did beautiful descriptive work. But where was the theory?'

Me: 'Yes? Where was it? I mean: you didn't go looking for it in Paris, London or New York where Lacanian psychoanalysis, actor-network theory or new materialism were thriving. These things were kind of lost on you.'

Monique: 'Yes, but for a good reason. My opportunity did not lay in convincing people working in completely different domains about the benefits of an actor approach. The opportunity was that I could make some inroads in areas of development sociology that where dominated by people who took for a fact that there existed some primary, fundamental mechanisms that could help explain larger scale social structures. But as crime writer and poet Dorothy Sayers put it, facts are like cows; if you look them in the face long enough, they generally run away. And this is precisely what I could lay bare by doggedly following the practices of the actors I engaged with.'

Me: 'Sounds pretty fundamental to me. But where did this all lead to?'

Monique: 'Well I couldn't say there's any clear-cut development of ideas here, but little by little the reasonable but subversive idea dawned on me that participants to social phenomena are as informed as outside investigators. Actually, they're way more informed. After all, actors *live in* the issues that we scholars think about for a limited time and from an external viewpoint. Downplaying actors' interpretations of situations just because these are not based on academic protocols is pretty arrogant, I would say. So, I quickly realised that, when it comes to herding, *ejidos*, or house evictions it was not me who was the expert, but rather the herders, the farmers and the activists I spent time with. They were far more experienced and knowledgeable of their situation than any sociologist or anthropologist could ever be. And from there it was a small step really to develop an interest in how, on the basis of their understandings, they exercised their political agency and savvy.'

Me: 'Political agency and savvy. These terms are a red thread in your work.'

Monique: 'Yes. They come from the field, so to speak, but they allowed me to hook onto the more interesting, more practical middle-range theories in development sociology and anthropology.'

[7] Luis González y González.
[8] El Colegio de Michoacán, Zamora, Mexico.

Me: 'You mean like Sally's idea of force fields.'

Monique: 'Yes, but also the work of Franz and Keebet von Benda-Beckmann. They were essential. First to carve out my niche in development sociology. And later to intervene in debates in legal and political anthropology. Terms such as 'corruption' or 'activism' look quite different you know when seen from an actor perspective…'

[Passenger noises as they pick up their stuff to prepare for the next stop]

Me: 'We're getting there. Time for one or two more questions.'

Monique *[resolutely]*: 'Shoot!'

Me: 'Most cherished actor-oriented concept?'

Monique: '*Agency*. Hands down.'

Me: 'Won't argue with you here. Most undervalued concept?'

Monique *[sighing]*: 'Interface.'

Me: 'Can you elaborate?'

Monique: 'Sure. Interface analysis is a handy methodological tool to zoom in on situations where discontinuities become apparent. But these discontinuities have been taken to suggest differences in lifeworlds, social fields or levels of social organisation in empirical settings of rural development. I see these discontinuities more between the *discourses of* those engaged in social intercourse in these empirical settings, and the *discourses on* these empirical situations by academics. It didn't occur to anybody that bridging these two discursive practices is important. I saw the importance of it, and it helped me embrace academic activism more and more. It links up with the whole debate about the performativity of the social sciences, you know…'

Me: 'Well I guess what makes you so special is that you have always made these realities talk to one another. You were always intent on confronting theory with actors' practices. Not like in the 'speaking truth to power' kind of way, but in a more humble, sensitising or reflective sense. And I'm sure that for a large part it is this that earned you the respect of academics in Zamora, Guadalajara, La Jolla, or Wageningen first, and later in wider European circles. This and your academic rigour. As Leo de Haan once told me when I asked him about how you were doing at Nijmegen[9]: 'Monique? Rock solid!'

[9] Radboud University Nijmegen, the Netherlands.

Monique: 'Nothing special really. Just a matter of doing your work meticulously. With passion. Academic life might be exacting, but it's also exciting!'

Me: 'And it has been so much more exciting with you around...'

[Smiles all over as the train prepares to stop]

Monique: 'This is my stop. I need to leave you here...'

Me: 'No problem. I get off at the next station. It was a pleasure, Monique. Thank you.'

Gerard Verschoor / Assistant Professor, Sociology of Development and Change Group, Wageningen University, the Netherlands / gerard.verschoor@wur.nl

20. FEARLESS INTELLECTUAL REINVENTION: MONIQUE NUIJTEN AND BEING AN EXPANSIVE ACADEMIC

Dennis Rodgers

I do not remember the first time I met Monique Nuijten – most likely over a drink (or two) at a NALACS or LASA conference, I imagine! – but I do remember when I first came across her work, and more specifically, her book, *Power, Community, and the State: The Political Anthropology of Organisation in Mexico* (London: Pluto, 2003). This was a reading for a development studies core course that I was teaching at the London School of Economics and Political Science in 2004, and it impressed me immensely when I read it. Effortlessly combining detailed empirical material with original and critical theoretical interpretation, the study offered a hugely engaging and thought-provoking vision of political processes that encompassed both individuals and organisations into a complex and contingent but coherent whole. As such, it bridged that most vexing of anthropological quandaries, how to reconcile micro and macro perspectives on the state and the exercise of power both within and against it. The study moreover did so in a context, Mexico, where the issue had long constituted a subject of significant controversy. Scholars had until then generally focused on the relative weight of phenomena such as clientelism, party politics, or institutional efficiency, and Monique's book grounded, holistic approach constituted a real breath of intellectual fresh air.

Power, Community, and the State remained a feature of my development studies reading lists for many years, until I moved from development studies (by then at the University of Manchester) to urban studies (at the University of Glasgow) in 2012. Although many of its theoretical insights and analyses remained relevant to social and political processes at work in urban contexts, its avowedly rural focus meant that it was not really an appropriate reading for my new teaching. I therefore assumed I would mainly encounter Monique and her work in the context of our mutual engagement with Latin American studies, most likely at LASA conferences. It therefore came as a real surprise when Monique wrote to me in 2013 to ask me to contribute an Afterword for a special issue on 'The Productive Spaces of Urban Development: Contestation and Creative Appropriation' that she was co-editing with Martijn Koster, with a view to submitting it to the journal *Urban Studies*. She explained that the special issue aimed to analyse how the enmeshing of formal urban planning encounters with informal modes of ordering gave rise to productive spaces in which new political configurations emerged, and that her contribution built on new ethnographic fieldwork that she had carried out in the slums of the city of Recife, in Brazil.

The vagaries of academic publication being what they are, the special issue ended up being published in the *Singapore Journal of Tropical Geography* (volume 37, issue 3) in 2016, re-titled 'Close Encounters: Ethnographies of the Coproduction of Space by the Urban Poor'. During the course of our exchanges about it, however, it became clear that the endeavour was part of a broader and remarkable process of wholesale reinvention of Monique's intellectual and professional trajectory, as she strove to transition from being a rural to an urban specialist. Such academic metamorphosis is relatively rare, but in addition, rather remarkably, it was not something that Monique engaged in only individually, but was in fact collectively leading. In 2014, she invited me to be a keynote speaker at a conference on 'The Urban Turn in Development Sociology: Agency and Ordering in the City' that she was organising at Wageningen University. Monique explained the logic behind the conference in the invitation email in the following way:

> *Originally our department was focused on rural development, but over the last decade we have slowly integrated more and more urban themes. For instance, I moved from land reform and peasant communities to slum upgrading and politics at the grassroots (Brazil and Spain). A colleague focuses on the urbanisation of the refugee question in East Africa. Another looks at the effects of post-conflict situations on cities (Latin America and Africa). Others look at the urbanisation of risks and hazardous cities. As you see, a broad range of themes. But as the focus on urban questions is quite new for our department/university, we have decided to organise a one-and-a-half-day workshop in which keynote speeches by urban development specialists will be combined with presentations by Wageningen researchers, to promote exchange and dialogue. In this way we hope to further establish the urban agenda of our group.*

This combination of being willing to fearlessly strike out beyond one's professional comfort zone, to seek to learn from the work of others with humility, and doing so moreover as part of collective project, is quite unique, and the workshop was a wonderful moment of intellectual exchange. The conversation was sustained subsequently, both virtually, as Monique would send me wonderful random emails about different urban topics that interested her – I particularly remember one asking for references about urban gangs and dancing! – as well as in person, after I moved to the University of Amsterdam in 2016. In this latter regard, while Monique's fearless and expansive intellectual reinvention is something that is clearly at the heart of her broader conception of 'what it means to be a "Good Academic" in the University today' – to borrow the title of a document that she co-authored in 2016 – it is also part of an ethos that is more than just intellectual. In the same document, she writes that being a good academic must also involve an *expansive* 'ethics of care' towards others, that should spill over into other aspects of life beyond the professional. In this regard, Monique brilliantly supported me during the two years that I navigated the idiosyncracies of Dutch academia, regularly inviting me to brunches – and in 2018 to watch World Cup football! – with her (and her partner Finn Stepputat) in Utrecht, in order to provide support and advice. Such behaviour on the one hand reflects her profoundly warm and generous personality, but on the other hand – in combination with her inclination to intellectual metamorphosis – is also part of a particular expansive academic ethos that Monique embodies, and which demarcates her as a remarkable and inspiring academic.

Dennis Rodgers / Research Professor, Department of Anthropology and Sociology, Graduate Institute of International and Development Studies, Geneva, Switzerland / dennis.rodgers@graduateinstitute.ch

21. ONTWIKKELEN DOE JE ZELF

Leontine Visser

Het is alweer bijna acht jaar geleden dat ik met pensioen ging en de vrijheid kreeg me met heel andere vormen van ontwikkeling bezig te houden. 'Ontwikkeling' heeft de dubbele betekenis van door derden gestuurde maatschappelijke veranderingsprocessen en individuele ontplooiing en bloei. Beide liggen idealiter in elkaars verlengde, maar zijn ook vaak strijdig.

Mijn vriendschap sinds 2015 met Fatma en haar drie kinderen geeft inzicht in het krachtenveld waarin een Eritrese vluchteling als alleenstaande moeder in Nederland zich staande moet zien te houden tegenover de onduidelijk gestelde brieven van gemeentelijke en andere overheidsinstellingen, artsen die te moeilijke woorden gebruiken en oudergesprekken op school. Maar dan is er die andere kant, die zo vaak onderbelicht blijft, van persoonlijke ontwikkeling als ontplooiing. Ik heb grote bewondering voor de groei die Fatma doormaakt, wanneer ze leert dat mannen in uniform niet je vijand hoeven te zijn, dat je ook anderen buiten je directe familie kunt vertrouwen, dat de onderwijzeres je kind gelukkig kan maken met wat het leert. En vooral, dat zij haar gevoel van eigenwaarde en zelfvertrouwen langzaam weer terugkrijgt – al duurt het jaren. En ik leer zoveel van haar over mijn eigen land!

Nu ik vrij ben van institutionele dwangbuizen, kan ik ontwikkeling op wetenschappelijk gebied dichterbij mijzelf houden en teruggaan naar mijn onderzoek in Indonesië uit de jaren 1980-90. Gedurende de laatste zes jaren heb ik een grote collectie beeldmateriaal beschreven met korte thematische teksten, en in drie talen boekjes geschreven bij de film die wij in 1982 maakten. Met mijn publicaties over Halmahera komt alles nu op een publieke website van het Museum van Wereldculturen in Leiden. Sinds 2015 ga ik weer ieder jaar naar Indonesië om vrienden op te zoeken en college te geven over mijn laatste (Indonesische) boek. Je kennis delen met studenten en met de oorspronkelijke eigenaren van mijn kennis over Halmahera is heel inspirerend. Met mijn grijze haren krijg ik bovendien nieuwe informatie, die mij destijds wegens mijn sociale positie als jongere vrouw onthouden werd.

Zo is ontwikkeling vooral een interactief proces, waarbij het niet alleen om ongelijke machtsverhoudingen en sociale positie gaat, maar vooral ook om het zoeken naar raakvlakken waar individuele en collectieve kracht versterkt kunnen worden. Dat geeft hoop.

Monique Nuijten liep voorop om sociale en politieke ontwikkeling te duiden als een 'hope generating machine' in de context van 'force fields'. Het was haar kracht om op te staan tegen maatschappelijk onrecht in Latijns Amerika en in Nederland, om de in een gesloten citeergemeenschap gevestigde definities en benaderingen kritisch te beschouwen en haar eigen weg te gaan. Hiermee heeft zij haar collega's in Wageningen en ver daarbuiten gestimuleerd en geïnspireerd. Maar deze laatste jaren werd Monique hard op zichzelf teruggeworpen en moet zij het hebben van haar eigen levenskracht in een onwelkom en deels onbekend krachtenveld. Intellectuele beschouwing verliest het dan van fysieke onmacht en emotionele spanning.

Maar je staat gelukkig temidden van het nu krachtigste sociale veld van je vrienden, dochters en kleinkinderen. Ik wens je het best mogelijke.

Leontine Visser / Professor Emeritus, Rural Development Sociology, Wageningen University, the Netherlands / voorhoevisserle@gmail.com

22. HOPE GENERATION AMONG REFUGEES AND MIGRANTS

Dorothea Hilhorst

A *liber amicorum* – a book of friendship – is a lovely occasion to revisit the early works of a friend. And how immensely pleasant it is to reread Monique Nuijten's book *Power, Community and the State: The Political Anthropology of Organisation in Mexico* on an idle rainy autumn afternoon, with the sun struggling to peek through the clouds, while the violent waves in the treetops visible from my window tell me I better stay inside. I initially read Monique's book prior to its prestigious publication by Pluto Press, when she defended her PhD thesis in 1998. Those were the years of youth, when two years' difference in age meant that one would truly look up to a colleague who had already 'made it' and defended her work so brilliantly and elegantly and with such composure that it entitled her to the distinction of *cum laude*.

Rereading the book in October 2020, the elegant brilliance of the book is still striking. It is also a book about a time gone by, when a major and legitimate notion of governance, in hindsight, represented the kind of fantasy that Monique made the subject of her analysis in the book. When Monique wrote *Power, Community and the State* (Nuijten, 2003), she could confidently cite authors arguing that we had entered a deterritorialised and transnationalised world – a world in which 'sovereignty has taken a new form, composed of a series of national and supranational organisms united under a single logic of rule,' as she quoted from Hardt and Negri (2000).

How dreamily naïve such a quotation sounds today. Today's world manifests itself as re-territorialised and re-nationalised, especially when seen through the eyes of migrants. Most passports in the world do not travel far. Borders that seemed to have dissipated have been reinstated as real borders, paper borders, iron borders, or even – when we read about the plans for barriers miles away from coastlines or hear of surveillant ships shooting at migrant boats at open sea – *borders of death*. 'The Human Rights Commission of the United Nations stated in 2018 that Europe has developed a refugee policy that implicitly and explicitly accepts death as an effective anti-migration instrument (Polman, 2020).'

Despite these developments, Monique's book is far from outdated. The core of her argument about the state as a hope-generating machine is more relevant than ever – certainly for the millions of migrants seeking entry to inaccessible states. The parallels with the Mexican peasants with whom Monique spent her ethnographic days are striking. Oliver Bakewell, for example, noted how potential migrants in East Africa are completely devoted to collecting papers and building a portfolio for an envisioned migration. During his presentation at the Forced Migration Studies Association Conference in Thessaloniki in 2018, Bakewell could have quoted Monique when saying that 'The culture of the state is central to the operation of the bureaucracy as a hope-generating machine. The hope-generating bureaucratic machine gives the message that everything is possible, that cases are never closed [...]' (p. 196). He could further have quoted her argument that 'State intervention in Mexico [migration policy in East Africa] tends to have a divisive effect on the population, and to frustrate independent collective organising efforts 'from below'' (p. 198). Rather than seeking out their brothers in fate and rising to protest, migrants are driven by the hope of becoming one of the lucky chosen few, and they do everything in their might to mould their behaviour and attitudes to the requirements imposed or favoured by the migration machines. The annual lottery that hands out 55,000 green cards to the United States – with chances of 1.33% for people in most eligible countries – is indeed the ultimate hope-generating machine, steering millions of people away from engaging in protests and activism, and motivating them instead to maintain immaculate track records and build their individual case files.

I am not the first – and probably will not be the last – scholar to recognise the relevance of Monique's notion of the state as a hope-generating machine for migration issues. The renowned migration scholar Heather Cabot drew on Monique's analysis in her 2010 ethnography of the pink card, the identity document issued to asylum seekers in Greece (Cabot, 2019). Cabot's paper analyses the bureaucracy around asylum, which is positive for a tiny number of people, whereas many more cases are rejected and an even greater number remain stuck in the system and never reach an outcome. The many faces and uses of the pink card result in a system where most applications simply disappear into the machine.

It is widely acknowledged that Europe's policies towards migration can be summarised by the word 'deterrence'. The European Union as well as its individual member states –

perhaps with the exception of Germany – seem united in their determined aggression in seeking to expose and render as visible as possible the cases of *failed migration* that result in an early death in the Mediterranean sea or being stuck in the horror of Moria and other camps. The purpose of these efforts is to deter would-be migrants from trying to reach Europe. Nonetheless, there are always a number of people who manage to slip through the cracks of the system and be granted asylum.

For a long time, I thought maintaining the appearance of a just system of asylum was a concession to the many Europeans who are supportive of refugees. In the Netherlands, for example, the government insists that there is no social support base for migrants. This, however, is far from the truth. Recent research from the University of Groningen (Kuppens *et al.*, 2019) found that, although the support base for migration is shrinking in the Netherlands, 45% of the population wants to support refugees. Another 25% of the population is open to supporting refugees provided that strict measures are taken to protect society from asylum seekers who misbehave. Thirty Dutch municipalities have declared their willingness to receive refugees from Moria. The bold statement of the right-wing Dutch government that there is no support base for refugees thus appears to be a malicious manipulation by a government that plays to the populist far right, where it fears losing its specific electorate. I always assumed that the few successful asylum cases were a triumph of the mass of refugee-friendly lawyers, volunteers, and left-wing politicians making noise on behalf of refugees. I assumed that they occasionally managed to beat the system.

Upon closer inspection, and after rereading Monique's book, I realise more clearly that those asylum seekers who successfully slip through the system are not a mistake or a failure of the deterrence machine. It is much more likely that the machine is built in such a way that, once in a while, a lucky individual comes out with a residence permit. It may very well be that the machine is designed in this way to discipline the migrants in Moria and other places, who are living a non-life, not unlike the shrivelled human heads on stakes that used to decorate the walls of medieval European cities to deter vagabonds from passing through the gates. Being told by their friendly lawyers to keep their heads down, behave well, and do whatever they can to enhance their chances in the procedure generates hope among asylum seekers. Knowing one or two people who succeeded before you further feeds that hope. Without it, refugees might unite and destroy the gates to Europe.

After this reflection on the relevance of Monique's work for migration studies, what remains is a big 'thank you'. I am glad to be able to contribute to this book of friendship for Monique. Everybody, no doubt, is a friend to some. With Monique, friendship defines her relationships with most if not all other people. Friendliness comes natural to her, yet is deliberate and sometimes even brave. I remember during my years at the Development Sociology Group in Wageningen, that Monique always maintained friendly relationships even if this meant going against the grain of the group. This made her friendship especially precious to me, and I felt immensely privileged when life granted us the serendipity of becoming neighbours in Utrecht as well.

Monique, I have so appreciated the many moments of our brief meetings, walks, coffee breaks on the terrace, and watching movies in the city centre. Thank you for your sincere friendship.

Dorothea Hilhorst / Professor of Humanitarian Studies, International Institute of Social Studies, Erasmus University Rotterdam, the Netherlands / hilhorst@iss.nl

References

Cabot, H., 2019. The European refugee crisis and humanitarian citizenship in Greece. Ethnos 84(5): 747-771.

Hardt, M. and Negri, A., 2000. Empire. Cambridge, MA: Harvard University Press.

Kuppens, T., Gootjes, F., Boendermaker, M., Gordijn, E. and Postmes, T., 2019. Ongenoegen, migratie, gastvrijheid en maatschappelijke onrust. Onderzoek Rijksuniversiteit Groningen, in opdracht van het Wetenschappelijk Onderzoek- en Documentatiecentrum. Available at: https://tinyurl.com/y6dxe9qk.

Nuijten, M.C.M., 2003. Power, community and the state: the political anthropology of organisation in Mexico. London, UK and Sterling, VA: Pluto Press.

Polman, L., 2020. Tegen elke prijs. Essay Vluchtelingen en Europa. Groene Amsterdammer. Available at: https://www.groene.nl/artikel/tegen-elke-prijs.

23. DE STUDIE VAN ONTWIKKELING; NIEUWE ENERGIE TE MIDDEN VAN UNIVERSITAIRE CRISIS ROND 1984

Philip Quarles van Ufford

Woord vooraf

In 1985 gaf ik een voordracht tijdens een seminar over onderzoeksplannen bij de vakgroep ontwikkelingssociologie in Wageningen. Tijdens de discussies viel me de hoge kwaliteit van de bijdragen van een aantal studenten op. Monique Nuijten was een van hen. Het leek wel alsof enkele studenten al getalenteerde jonge collega waren van hun leermeester, Norman Long. Hij werkte met hen al samen, zo bleek. Deze samenwerking verschafte het vakgebied nieuwe energie. Van deze groep studenten ging ook op langere termijn veel kracht uit.

In die tijd ontstond op het terrein van de ontwikkelingsstudies een vrij hechte samenwerking tussen 'Wageningen' en de Vrije Universiteit Amsterdam (VU). Ook met de studente Monique Nuijten. Vrij snel na genoemd seminar vroeg ik haar om in Amsterdam een gastcollege te komen geven. De samenwerking was zeer inspirerend.

In deze bijdrage reflecteer ik op de ingrijpende transformatie van de landelijke agenda van ontwikkelingsgericht onderzoek. Deze vernieuwing begon aan het begin van de jaren tachtig. Na 1982 diende zich ook een diepe crisis aan in het gehele universitaire bestel. De overheid kondigde zeer ingrijpende bezuinigingen aan. De schok was enorm. Het was alsof een wolk van zorg en onzekerheid over de universiteiten trok. Hoe moest het verder? Alles was in principe onzeker geworden.

Over de universitaire crisis

Nationaal 'beleid' kreeg in het begin van de jaren tachtig een sterke greep op het universitaire bestel. Het voortbestaan van vakgroepen, ja zelfs van universiteiten, werd onzeker. Die werd nog groter omdat de overheid niet duidelijke contouren van een nieuw inhoudelijk beleid liet zien. Deze onzekerheid dreigde velen te verlammen. Ieder tastte in het duister. Eigen nadenken leek geen zin te hebben; het angstig afwachten maakte velen murw. Overleven, hoe dan ook, daar ging het nu om. Een universitair bestuurder drukte het zo uit bij een eerste overleg over nieuwe onderzoeksplannen:

> *het gaat er nu niet meer om wat wij zelf willen. We moeten ons nu niet meer op de toekomst richten vanuit onze eigen kracht en eigen voorkeuren. Nee, we moeten ontdekken welke verwachtingen en prioriteiten over een paar jaar op ons afkomen. Daarop moeten we inspelen in onze planning. Alleen zo kunnen we overleven.*

Maar wisten we wel wat er van ons verwacht zou gaan worden? Dat was namelijk onduidelijk. Moesten we dan voor alles openstaan? Opende dit niet de weg naar een nieuw soort nihilisme; naar een nieuw soort markt van 'U vraagt en wij leveren'? Gaven we zo het eigen denken over de wetenschappelijke toekomst niet uit handen? Zou dit ons op langere termijn dan niet schaden? Toch kreeg deze overlevingsstrategie her en der de overhand. De relatieve autonomie van facultaire sturing verminderde. En colleges van bestuur konden dit tij niet keren. Hun traditionele rol als facilitator van facultair beleid verdween. Zij werden vooral uitvoerder van regeringspolitiek. Langzamerhand kondigde zich een bedrijfsmatige aanpak aan, met producten, vraag en aanbod.

De onzekerheid verlamde velen. Plotseling vertoonden zich allerlei 'waarzeggers' die claimden wel te weten wat de bedoelingen van de overheid waren en dus van de gevraagde producten. Een van hen vertelde mij later hoe verbaasd hij was geweest dat zovelen zijn verhalen zomaar hadden geloofd. De grens tussen waarzeggerij en managen was maar heel dun. Van post-modern denken was nog geen sprake, al is dit vaak gezien als veroorzaker van het 'nieuwe publieke management'. Deze relativerende wending in het wetenschappelijk denken kreeg pas aan het eind van de jaren tachtig invloed. Het post-modernisme was niet de oorzaak van deze crisis. Veeleer was het een gevolg van enorme bezuinigingen. In deze context veranderde ook de agenda van ontwikkelingsstudies. Maar het liep anders.

En de ontwikkelingssociologie/antropologie?

In het grote bezuinigingsgeweld speelden de afdelingen ontwikkelingssociologie en –antropologie maar een kleine rol. Veel geld viel daar niet te halen. Merkwaardigerwijs was daar geen sprake van verlamming. Maakte de marginaliteit van dit vakgebied misschien een zekere autonomie mogelijk? Het leek wel alsof er van een diepe crisis geen sprake was. De zoektocht naar vernieuwing van de agenda van de studie van ontwikkeling was al eerder op gang gekomen en ging tijdens de crisis verder. Dit leidde in 1984 tot een landelijke conferentie waarin een fundamentele herijking van lang gevestigde agenda's werd gepresenteerd. Dit was ingrijpend. Er werd afstand genomen van ideologische preoccupaties. Vele Nederlandse ontwikkelingssociologen en -antropologen bleken in staat om oude visies kritisch te bezien, oud gelijk te relativeren. De bereidheid groeide om met oude 'tegenstanders' samen te werken. De vernieuwing sloeg aan. Er ontstond een nieuwe consensus: de politisering van het vak was te ver doorgeschoten.

Nu werd het mogelijk om gezamenlijk vorm te geven aan een nieuwe agenda. Het boek 'Trends en tradities in de ontwikkelingssociologie' (Husken *et al.*, 1984) doet tussentijds verslag van deze herijking. Niet lang daarna werd ook een breed nieuw nationaal onderzoeksprogramma gepresenteerd. Het roer ging om.

Dit nieuwe onderzoeksplan kreeg de naam 'Schakelingen'. Dit woord drukte de gewonnen vernieuwing goed uit. Relaties tussen verschillende actoren in het ontwikkelingsproces werden geproblematiseerd, waarover tevoren vooral politieke zekerheden opgeld hadden gedaan. Oude zekerheden en veronderstellingen van samenhang en systematiek over deze schakelingen werden opengebroken. Het vakgebied van ontwikkelingsstudies ontworstelde zich zo aan de verenging van 'toegepaste' wetenschap. Toepassingen werden nu zelf voorwerp van kritische studie.

Deze nieuwe vitaliteit trok na enige tijd ook de aandacht van de overheid. Het Ministerie van Onderwijs reageerde met waardering op het nieuwe onderzoeksprogramma. Nogal onverwacht zegde de overheid zelfs aanzienlijke nieuwe fondsen toe. Dit gaf het initiatief nieuwe kracht. De levensvatbaarheid van de nieuwe plannen bleek heel duidelijk toen het Ministerie van Onderwijs haar financiële toezeggingen plots bleek te zijn vergeten. De herwonnen vitaliteit ging toen niet verloren. Het nieuwe plan bleek echt levensvatbaar. Men ging aan de gang.

Hoe is dit onverwachte succes te duiden? Speelde de marginaliteit van ontwikkelingsstudies in het hoger onderwijs mogelijk een rol? Het leek wel alsof de crisis aan het vakgebied voorbij was gegaan. Korte tijd later kwam het ook tot een bloeiende samenwerking met Duitse en Britse collega's in het Europese EIDOS programma, dat gedurende bijna twintig jaar grote vitaliteit toonde in een reeks van conferenties en publicaties. De min of meer autonome vernieuwing van de agenda van ontwikkelingsstudies bleek over een langere periode vruchten af te werpen, nationaal en internationaal.

Monique Nuijten en een groep geëngageerde Wageningse studenten

Deze vernieuwing kreeg een krachtige impuls van een groep getalenteerde en geïnspireerde studenten uit Wageningen. Monique Nuijten nam onder hen een bijzondere plaats in. Deze studenten – waren ze derdejaars? – betoonden zich tijdens discussies in seminars en daarbuiten vrijwel collega's van hun eigen hoogleraar, Norman Long. Zij bouwden voort op zijn inzichten en hielpen zijn nieuwe visie op ontwikkeling in onderzoeksplannen uit te werken.

Ik ontdekte dit toen de samenwerking tussen Wageningen en de VU werd opgebouwd. Het engagement van deze studenten viel me op. Zij bleken ook getalenteerd. Zo maakten het analytische vermogen van enkelen, onder wie Monique Nuijten, grote indruk op mij. Deze studenten werden inspirerende dragers van een nieuwe onderzoek agenda. Zo voegden ook zij inhoudelijke vitaliteit en werkvreugde toe aan het landelijke 'Schakelingen' programma. Aan bredere kaders van samenwerking ontleenden deze studenten inspiratie. Zij voegden daaraan hun energie en kwaliteiten toe. Eerst als student en later als onderdeel van de wetenschappelijke staf. En daar inspireerden zij weer anderen.

Om welke agenda ging het bij de samenwerking tussen Wageningen en de VU? Norman Long stelde het begrip 'lokale ontwikkeling' centraal. Hij had eerder het primaat van nationale en internationale ontwikkelingen krachtig gerelativeerd. De relatieve autonomie van lokale ontwikkelingen kreeg nu zijn volle aandacht. Dit leidde tot zicht op de grote diversiteit in lokale verandering. Dit maakte weer interessante vergelijkende studies mogelijk.

Aan de VU was een vergelijkbare verandering opgetreden. Daar werd het concept 'ontwikkelingsbeleid' opengebroken. De assumptie dat de verschillende praktijken van 'beleid' een systematisch en rationeel geheel vormen werd onderzocht. Dit leidde tot verrassende inzichten. Van een veronderstelde samenhang en rationaliteit van een aantal praktijken in een 'beleid' bleek niet of nauwelijks sprake. Heeft het woord beleid in de internationale ontwikkelingsproblematiek misschien vooral politieke en legitimerende betekenis? En belemmert dit concept van beleid misschien juist niet een beter begrip van een reeks van gebeurtenissen die op elkaar inwerken? Er valt nog veel te doen.

Monique en een groot aantal collega's, her en der, gingen jarenlang geïnspireerd aan de gang. Zij hielpen met het leggen van een stevige basis voor nieuwe ideeën.

Coda: leren deze ervaringen ons iets? Nog een aantal opmerkingen

Deze geschiedenis ligt alweer 35 jaar achter ons. Vele onderzoekers die toen geïnspireerd begonnen, komen nu aan het einde van hun werkend leven. Nieuwe collega's en nieuwe agenda's dienen zich aan. Ook het universitaire bestel is in de afgelopen decennia ingrijpend veranderd. Inzicht in deze gebeurtenissen en veranderingen is toch van belang. Ook al lijkt alles wel anders te zijn geworden, we kunnen we nog wel iets leren van het verleden. Met enkele opmerkingen daarover wil ik besluiten.

1. Over erkenning van eigen onwetendheid
   Een belangrijke aanzet tot vernieuwing lag in 1984 in de bereidheid om eigen tekorten openlijk onder ogen te zien. De afschaffing van ideologische zekerheden maakte nieuwe leerprocessen mogelijk. Onwetendheid kwam in een ander licht te staan; de betekenis ervan veranderde. Erkend falen werd prikkel voor vernieuwing. Deze onwetendheid betrof vooral het verleden van de vaste universitaire staf. De bereidheid om nieuwe vormen van samenwerking aan te gaan gaf ook aan getalenteerde studenten nieuwe kansen. We zagen dat in Wageningen. Een aantal van hen, onder wie Monique Nuijten, greep deze kans met beide handen aan.

2. Een bestel gebouwd op wantrouwen schiet tekort als basis voor nieuw universitair onderzoek
   De ingrijpende vernieuwing was rond 1985 mogelijk, omdat onderzoekers zich nog een zekere ruimte en autonomie konden verwierven, zelfs in crisistijd. Deze ruimte is vrijwel verdwenen. NWO heeft vooral een smorende werking. Onderzoekers worden niet meer vertrouwd. Zij dreigen te verzanden in een reeks van dwaalwegen die we 'kwaliteitszorg' zijn gaan noemen. Het fundament van dit wantrouwen werd verstevigd in de zogenaamde bedrijfsmatige aanpak. Resultaten van onderzoek werden gezien als een product van onderlinge competitie, en van beteugeling van kosten. De controle over de 'vraag naar onderzoek' werd toegewezen aan financiers als NWO. De zoektocht naar geld werd een teken van innovatieve slagvaardigheid.
   Defensieve preoccupatie met overleven liep uit op een bestel dat eigen ruimte afbrak. Ruimte voor een eigen blikveld was klein geworden. Kwaltiteitszorg ligt nu vooral bij anderen. Wanneer onderzoekers zelf een nieuwe richting willen inslaan, dan zijn de officiële hordes hoger dan ooit. Een tamelijk verlammend nihilisme kenmerkt thans het bestel.

3. Nieuw wetenschappelijk onderzoek naar nieuwe globale problemen.
   In de jaren tachtig leverde de bereidheid om ontwikkelingsgericht onderzoek te *depolitiseren* nieuwe energie op. Rond 2020 is vooral een proces van *re-politisering* van de universiteiten wenselijk. Grote nieuwe problemen dienen zich aan: de klimaatcrisis en twijfel aan de maakbaarheid van veranderingen. De twee probleemvelden zijn met elkaar verbonden. En er is al enig werk verricht.

Krijgt nieuw onafhankelijk onderzoek in het verlamde bestel een kans?

Maar nu nieuwe globale crises zich aandienen vallen deze tekorten nog meer op. Recent schreven de 'Wageningers' Büscher en Fletcher (2020) een belangrijk boek over de klimaatcrisis. Zij stellen dat voor een goede analyse en aanpak een echte *revolutie* nodig is! Maar wat voor revolutie? Mij dunkt een revolutie vlakbij huis: de universiteit. Onderzoekers moeten nieuwe ruimte krijgen. Zij worden nu vrijwel totaal ingesnoerd en staan voor veel zwaardere taak dan in de jaren tachtig. Leidt het wantrouwend bestel er niet toe dat velen en vaak juist besten snel afbranden? Het in de organisatie vastgezette wantrouwen vormt een groot gevaar. De weg vooral gericht op overleven is een dode weg.

Waarom pakken Colleges van Bestuur dit nieuwe politieke probleem niet aan? Is het niet tijd dat zij eigen regie gaan voeren? Misschien moeten zij gaan erkennen zelf vooral onwetende studenten te zijn. De grenzen tussen wetenschap en politiek kunnen opnieuw worden gedefinieerd. Universiteiten moeten daartoe zelf politiek actief worden. Dat zou inderdaad 'revolutie' zijn,

Daarbij is het goed onze assumpties over de relatie tussen wetenschap en politieke en maatschappelijke organisaties kritisch te herzien. Deze waren nogal naïef. De bereidheid om serieus aandacht te geven aan resultaten van onderzoek blijkt vaak nogal problematisch. De behoefte om *niet* te weten kan buitengewoon krachtig zijn. Meestal is er sprake van een gevoel van grote machteloosheid bij politieke organen. Zeker, kennis ontmoet soms macht, maar ook heel vaak machteloosheid. En misschien is het onderscheid tussen de twee ook nog problematisch. Moeten universiteiten niet creatief op zoek naar nieuwe politieke relatievelden. Een klein voorbeeld moge het belang ervan verduidelijken.

Toen de Inspectie te Velde van het Ministerie van Buitenlandse Zaken haar 25-jarige verjaardag in 2002 vierde op Clingendael, vroeg de directeur van de Inspectie te Velde aan het einde van zijn referaat aan de Minister waarom zij, net als haar voorgangers, wel veel waardering uitsprak voor de kwaliteit van het evaluatiewerk, maar er verder nauwelijks aandacht aan gaf bij de vorming van nieuw beleid. De Minister gaf dit openlijk toe. Zij legde ook uit waarom dit zo was. Was het niet haar primaire taak om de begroting door de Tweede Kamer te krijgen? De prioriteiten daar moesten we de grootste aandacht krijgen om haar beleid te kunnen opzetten en uitvoeren. Haar advies aan de directeur van de Inspectie was dan ook: richt je met je studies primair op het parlement. Ik kan er gewoon niet veel mee. Zij betoonde in die erkenning van eigen machteloosheid politieke moed. Er veranderde overigens niet veel. Kunnen we daarvan alsnog iets leren?

Colleges van Bestuur kunnen hun eigen grenzen van machteloosheid nog eens onder ogen zien en de relatie met de politieke en maatschappelijke omgeving herijken. Een intelligente re-politisering moet mogelijk zijn, zeker als de belangen van goed onderzoek zo groot zijn geworden. Je moet wel weer een toegewijde student durven worden. Zoals Monique, in de beste betekenis van het woord, altijd is gebleven.

Philip Quarles van Ufford / Emeritus, Anthropology of Development, Vrije Universiteit Amsterdam, the Netherlands / flipq@dds.nl

Referenties

Büscher, B. and Fletcher, R., 2020. The conservation revolution: radical ideas for saving nature beyond the anthropocene. London: Verso.

Hüsken, F., Kruijt, D. and Quarles van Ufford, P. (eds), 1984. Trends en tradities in de ontwikkelingssociologie. Muiderberg: Coutinho.

24. FEAR AND FANTASY IN SOCIAL MOVEMENTS – AND IN THE RESEARCH ABOUT THEM

Stine Krøijer

In December 2015, I received an email from Monique Nuijten with an invitation to participate in a panel in the POLLEN16 conference in Wageningen on social movements and political activism at the grassroots. A few days before the email landed in my inbox, I had been teaching her article *Between Fear and Fantasy: Governmentality and the Working of Power in Mexico* (2004) to a group of graduate students in Copenhagen. The mere sight of an email coming from Monique immediately produced a flash of fear: Did the email contain a complaint about how I had shared her article with students (my failure to observe property rights and the usual University bureaucracy)? Or had I grossly misinterpreted her insightful piece about rural peasants in Mexico and the fears and fantasies produced in their encounter with state bureaucracies? My fears luckily proved to be completely unjustified and our relationship soon developed into a fruitful conversation about political activism and the imaginative capacities at play in social movements in Europe.

A few years before our first encounter at the conference in Wageningen, Monique had shifted her research interests from peasant-state relations in Mexico via expressions of citizenship in Brazil to political activism in Spain in the aftermath of the financial crisis. She was working with the Plataforma de los Afectados por la Hipoteca (PAH), a network of mortgage victims and activists who were fighting against the eviction of poor homeowners who were no longer able to pay their debts after the crash of the housing market. In the years before the financial crisis, banks had seduced low-income groups with promises of private homeownership and social respectability associated with, but now they found themselves trapped by flexible loans and unintelligible mortgage bonds. Building on the experiences from the Indignados and the broader anti-austerity movement in Spain – also referred to as M-15 due to its occupation of the Puerta del Sol square in Madrid on May 15th, 2011 – PAH strove to change the bleak outlook to the future of these permanently indebted families.

The platform employed horizontal organising and a mix of direct action tactics and the making of legal proposals to curb the financial speculation of banks, to initiate schemes for debt cancellation and secure dignified housing for all. In this context, Monique described 'the political' as an antagonistic social arena and was, in continuation of her earlier work, concerned with describing the political agency of people at the margins and how it comes into being at an interface with existing structures of power (Nuijten, 2015). In her limited published work on the PAH, Monique shows how the members of the platform constituted themselves as political subjects through narratives. Individual life stories of mortgage victims told at the weekly open meetings of the platform and stories about politicians, banks and the state committing injustices or failing to protect and work in the interest of common people served to make sense of their place in the world and to form political strategies (ibid. 484).

Anchored in an admirable commitment to ethnographic analysis of people's everyday lives and mundane interactions with the state, Monique continues the early work of Norman Long (1992), Franz and Keebet von Benda-Beckmann (2013) and others in critical development studies. Her work falls at the interface of political and legal anthropology and carefully illuminates how political power plays out in practice. Law, corruption and citizenship are not preexisting entities, but come into being in social interaction (Lazar and Nuijten, 2013; Nuijten, 2013); a process that also involves a work of imagination. In the case of the PAH she shows how storytelling became a way for the mortgage victims to engage with the crisis and constitute themselves as political subjects by mobilising moral notions of justice out of joint with those held by corporate bankers and their associated yes-men.

In the everyday practices of the movement, the narrative contestations of austerity politics were closely associated with public performances of dissent. Mortgage victims engaged in collective resistance outside homes where families were facing eviction by the police organised so-called *escratches*, where movement participants would go into the intimidating bank lobbies to stage a conflict or take the accusations to the responsible politicians or bankers' own homes. Such performances of dissent work for people to see themselves as political subjects, which in practice also entails overcoming one's fears of the likely repercussions and the need to bracket one's embarrassments about being poor in order to take action. To become a political subject in this context very likely also entails the capacity to hope and believe that things could be different.

It is probably around the power of the imagination and its role for the formation of political subjectivities that I feel the strongest association with Monique's work. At the time of our joint panel about grassroot activism in 2016, my own work was focused on the growing climate justice movement in northern Europe and its continued experimentation with forms of action, particularly in relation to the fossil fuels industry and its insistence on continuing the expansion of lignite mining in Germany. Where Monique's paper addressed the difficult relationship between affected families and movement activists, my own paper concerned the mobilisation of nature (trees) in webs of mutual political aid around mining conflicts. The concern about the climate, which led to practical arrangements with trees through tree sitting, was enabling the formation of more-than-human political subjectivities, I suggested (Krøijer, 2019). Despite our differences and because of Monique's superior experience, the encounter resulted in a rich and ongoing conversation about different ecologies of political contestation, how new forms of activism in Europe question the individualisation of responsibility (for climate change, financial crisis) and about the imaginations that activists tend to hold of their world.

As our relationship grew closer, it also involved Monique's gentle mentoring about how to handle the fears and fantasies associated with academic life, which I appreciated immensely. From early on, Monique added an attention to people's fantasies about the state to the more strictly interactive focus of critical development studies at Wageningen. Her work illuminates how people tend to imagine the state as a distinct and unitary entity with a mysterious agency and an almost omnipotent ability to pull the strings (Nuijten, 2004; Taussig, 1992). My own work had shown how left radical activists envisioned capitalism as an all-encompassing system with the power to absorb and coopt any resistance, offering little chance of escape, let alone for radical transformation of the relations of power (Krøijer, 2015). Such gruesome imaginaries of powerful opponents inevitably shaped activist narratives and forms of actions, but also another kind of 'fantasies' about future scenarios. Much the same about fantasies, omnipotence and alternative futures could probably be said about significant parts of academic life.

Yet fantasies also hold a more hopeful dimension. Over the following two years, I had the pleasure to return to Wageningen to participate in a discussion about social movements between engaged scholars, activist researchers, people that work in NGOs and political activists. The conversation concerned the ways in which the political character of movements – such as the platform of mortgage victims in Spain, climate justice activists in Germany, and citizen groups opposing shale gas fracking in the Netherlands – place researchers and their knowledge production in a political place as well. One the one hand, movements are actively involved in producing knowledge to inform their political struggles, in order to gain credibility in political decision-making processes and to transform relations of power (Roth *et al.*, in press), and on the other hand academic knowledge about movements circulates and becomes reappropriated by movement actors themselves. A hopeful dimension is luckily inevitable in collaborative processes due the promise they hold for contributing new knowledge to political processes, although this also involves sensitive issues about how to manage one's own position and sympathies in movement research. Like other participants in the room, I had made a journey from trying to hold my worlds and engagements separate, to recognising that collaboration does not (necessarily) stand in the way for critical analysis. Participation in knowledge generation about the future effects of natural resource exploitation in a specific area, for example, does not prevent the possibility of also understanding the power of such predictions.

In sum, the question of knowledge production in and about social movements reiterates how fears and fantasies are intrinsic to movements – as it probably also is to academic life and life in general. Imagination about others and about the future entails the outline of everything from pleasant possibilities, the contemplation of improbable and undesirable scenarios to dystopic visions, often generated by extrapolating from trends of our time (Krøijer, 2020). In fantasy as a genre of fiction, this imaginative capacity is employed to create alternative worlds where the struggle between good and evil is finally settled. Activists in social movements are like most of us, torn between fears about what is yet to come and fantasies about alternative, better futures. As Monique Nuijten has pointed out, this is the essence of hope.

Stine Krøijer / Associate Professor in Anthropology, University of Copenhagen, Denmark / stine.kroijer@anthro.ku.dk

References

Von Brenda-Beckmann, F. and Von Brenda-Beckmann, K. (2013) Political and legal transformations of an Indonesian polity: the Nagari from colonisation to decentralisation. Cambridge: Cambridge University Press.

Krøijer S., 2020. Civilization as the undesired world: primitivists and indigenous peoples in times of climate crisis. Social Analysis: The International Journal of Anthropology 64(3): 48-67.

Krøijer, S., 2019. Slow rupture: the art of sneaking in an occupied forest. In: Holbraad, M., Kapferer, B. and Selma, J. (eds) Ruptures: anthropologies of discontinuities in times of turmoil. London: UCL Press.

Krøijer S., 2015. Figurations of the future: forms and temporalities of left radical politics in Northern Europe. Oxford and New York: Berghahn Books.

Lazar, S. and Nuijten, M., 2013. Citizenship, the self and political agency. Critique of Anthropology 33(1): 3-7.

Long, N., 1992. Battlefields of knowledge: the interlocking of theory and practice in social research and development. London: Routledge.

Nuijten, M., 2004. Between fear and fantasy in governmentality and the working of power in Mexico. Critique of Anthropology 24: 209-230.

Nuijten, M., 2013. The perversity of the 'Citizenship Game': slum-upgrading in the urban periphery of Recife, Brazil. Critique of Anthropology 33(1): 8-25.

Nuijten, M., 2015. Political activism of the PAH (Platform of Mortgage Victims) in Spain: the significance of Franz von Benda-Beckmann's work for the study of power, political agency and legal pluralism at the grassroots. The Journal of Legal Pluralism and Unofficial Law 47(3): 476-492.

Roth, D., Köhne, M., Rasch, E. and Winnubst, M., 2020. After the facts: producing, using and contesting knowledge in two spatial environmental conflicts. Planning, Politics and Space in press. https://doi.org/10.1177/2399654420941513

Taussig, M., 1992. The nervous system. London: Routledge.

25. PUSHING RESISTANCE AND REPRESENTATION: OPENING PATHWAYS FOR SCHOLAR-ACTIVISTS

Jessica Duncan

In one of her co-authored papers, Monique Nuijten wrote that 'utopian aspirations are part and parcel of subcultures of resistance' (Bode Bakker and Nuijten, 2018: 213). This ambition to seek out ideals through creative resistance embodies Nuijten, as a scholar, a colleague and a friend.

In this short essay I reflect on three aspects of Nuijten's scholarship: citizenship as a way of side-lining the most affected; scholar-activism as approach and response; and, seeking out the stability of informal (research) spaces. I reflect on these in relation to not only my own career trajectory, but also the current state of academia, before concluding with a reflection on the value and importance of female scholar activists, in particular at Wageningen University.

The citizenship agenda

Nuijten's work around the 'citizenship agenda' stressed how the logic of 'responsible citizens' coupled with empty participatory procedures in development projects have the perverse effects of side-lining the poor and reinforcing clientelist politics (Nuijten, 2013). Nuijten's work on citizenship thus offers us a way of framing anthropological enquiry into politics, also in political terms. In her work, she has asked how can citizenship be experienced in any given context, so as to explore how particular political communities and political agency are constituted (Lazar and Nuijten, 2013).

A preoccupation of my own work has been to understand how the actors of diverse food sovereignty movements exert political agency in spaces that have traditionally excluded them (Claeys and Duncan, 2019; Duncan and Claeys, 2018). I have documented and analysed how food sovereignty movements resist this side-lining Nuijten describes, and how they work to reinforce the voices of the 'most affected'. More specifically, I have examined internal movement procedures developed to strengthen and protect the voices of food producers and associated political identities when converging with other movements, and when engaging in more formal policy spaces. Using anthropological methods to explore relational politics, I have been able to show how movements are re-politicising spaces that have been subjected to decades of depoliticisation (e.g. food security governance and policy spaces) (Duncan and Claeys, 2018).

These, and other questions of citizenship, and the participatory turn in policy and development, required ongoing, critical, in-depth qualitative analysis building on Nuijten's approach and resulting insights. Doing so also requires a commitment to the people and movements under study: said otherwise, scholar-activism.

Scholar-as-activist

Activist scholarship aims to (co)produce knowledge through active engagement with, in and/or for social movements with a view towards breaking down dominant hierarchies of knowledge and power. Such work pays careful attention to the complexity and taken-for-grantedness of lived experiences and social context. As a scholar-activist, Nuijten embodies this approach and has pushed boundaries and minds through publications, education and public engagement. Her commitment to emancipatory scholarship has certainly supported the lives of many people, particularly in Brazil, Mexico and Spain. Her publication *Lucha y Esperanza: Testimonios de participants en la PAH* (Struggle and hope: testimonies of participants in the platform of people affected by mortgages) (Nuijten *et al.*, 2019), in particular stands out. When she handed me this booklet to read as part of our collective work as the Political Agency at the Grassroots Cluster of the Centre for Space, Place and Society, she grinned. It was a smile that conveyed that she was proud of the work, but also that she could not believe she had managed to get it printed. She explained that she was still in shock that the university publisher actually printed such content! The booklet is an excellent example of what working with and for movements can look like. The output is a recognition of peoples' struggles and experiences, and a validation of their knowledge. In reading her work, engaging with her and her ideas, and learning from the way she walks the fuzzy line that scholar-activism demands, Nuijten's approach has directly influenced my own work as a scholar-activist, and certainly many more.

And this is critical, because there is a current urgency for scholar-activism. Science and scholarship are under threat: legitimacy and trust must be re-built at various levels of society. And doing so may, in many instances, mean working against the rationale of our academic institutions. Around the world, universities are 'experiencing a crisis of both purpose and form' and they shift towards a focus 'narrowly on serving the needs of the market' (Ross, Savage, and Watson 2020). As argued by Ber and Seeber (2016), the corporatisation of the contemporary university has intensified workload, demanding increased speed and efficiency from faculty regardless of the consequences for education and scholarship. Reinforcing this point, Borras (2016: 36) contends that:

> *reward and punishment are powerful formal and informal norms and rules inside the academy and social movements, partly as an instrument of accountability, that largely determine what scholar-activists can do and cannot do. Inside academic institutions, what is rewarded are works that are deemed to have academic rigor translated to actual publication outputs that in turn make significant impact based on academic criteria.*

Nuijten's analysis of citizenship as a perverse logic to silence, can be helpful for analysing the logic on our universities and the implications this has for change. Our endless requests to participate in 'bottom-up' university processes only to be presented with finalised agendas, completed action plans and restricted debate in the spirit of constructive consensus and a unified vision only reinforces this point. Further, while many efforts are underway challenging the formal rules and norms, Nuijten reminds us of the importance and value of targeting the informal.

The stability of the informal

In her work on urban planning, Nuijten usefully contests the idea of 'the informal'. She argues that the informal and formal should be analysed as interconnected (i.e. relationally). Rejecting the tendency to conflate the formal with the tree and the informal with the rhizome, Koster and Nuijten (2016) argue that the formal tends to be arbitrary and frightening (rhizomatic), whereas informal practices can be very predictable and stable (arboreal). Based on this, they called for a re-valuation of the informal sphere.

My own experience as a scholar-activist support these insights. For my doctoral research, I was immersed at the crossroads of formal international diplomacy and movement mobilisation. It was in the daily informal, taken-for-granted practices that the stability and predictability of governance systems were uncovered, for example, the hegemony of neoliberal, or more specifically in my case, sexism. Considering the current state of academia, and drawing inspiration from Nuijten, I am increasingly convinced that we need to unearth and make visible the many informal, subtle practices that structure our working lives, including institutional sexism, racism, and classism. Their stability is indeed what needs to be cut away. In practice, this means not only working to change formal structures through committee work, letters, petitions, manifestos and strikes for example, but also identifying, calling-out, and dismantling the myriad structures of injustice that operate as hidden arboreal root system holding together the status quo that is, from where I am standing, failing.

And at the same time, and at risk of contradicting myself, as a scholar-activist, much like Nuijten has, I draw inspiration from these informal spaces. Amidst the ever-changing expectations being placed on academics, and recognising a tendency here to romanticise things, the grounded nature of movements, the concreteness of their demands, the urgency of timelines and actions, the passion and conviction, these are what give energy, inspiration and motivation.

Representation in academia

An editorial in the scientific flagship journal *Nature*, confirms what many social scientists have long acknowledged: 'Science and politics have always depended on each other' (Anon, 2020). Another *Nature* comment highlights how researchers are avoiding working with activists [on racial justice] out of worry that an 'activist' label could have negative implications for their careers (Bassett, 2020). Basset (2020) argues that this is typically self-censorship, enforced by norms of 'professional' behaviour'. But recognising politics and power, and explicitly doing politics through research, is not one and the same. Further, doing research for emancipatory politics, in solidarity with the struggles, the fights, the tears, the laughter, the songs and the victories of people, is not self-evident or easy.

Representation matters, and this extends to scholarship. Having a woman, in this case Nuijten, thrive in a competitive, male-dominated, patriarchal environment, and successfully work as a scholar-activist, has opened up pathways for scholars such as myself. Regardless of one's commitment and politics, working against your own job security, or at least the logic of the institution that employs you, is at some point guaranteed to be unnerving. As an early career scholar, trying to find my way in a university, one that is often perceived to be at odds with many of the conclusions I have come to in my own research, I drew reassurance knowing that there were others, blazing the trail ahead of me.

As trust in science and scholarship declines, and as the challenges we face becomes increasingly complex, there is an urgent need for activist-scholarship that is aligned with the experiences and knowledge of those fighting for their rights. Let us strengthen the subcultures of resistance to move towards a more just future. Nuijten's work has many of the key ingredients to make this happen.

Jessica Duncan / Rural Sociology Group, Wageningen University, the Netherlands / jessica.duncan@wur.nl

References

Anonymous, 2020. Why nature needs to cover politics now more than ever. Nature 586(7828): 169-170.

Bassett, M.T., 2020. Tired of science being ignored? Get political. Nature 586(7829): 337-337.

Berg, M. and Seeber, B.K., 2016. The slow professor: challenging the culture of speed in the academy. Toronto: University of Toronto Press.

Bode Bakker, M. and Nuijten, M., 2018. 'When breaking you make your soul dance' utopian aspirations and subjective transformation in breakdance.' Identities 25(2): 210-227.

Borras, S.M., 2016. Land politics, agrarian movements and scholar-activism. Inaugural Lecture. The Hague: ISS.

Claeys, P. and Duncan, J., 2019. Do we need to categorize it? Reflections on constituencies and quotas as tools for negotiating difference in the global food sovereignty convergence space. Journal of Peasant Studies 46(7): 1477-1498.

Duncan, J. and Claeys, P., 2018. Politicizing food security governance through participation: opportunities and opposition. Food Security 10(6): 1411-1424.

Koster, M. and Nuijten, M., 2016. Coproducing urban space: rethinking the formal/informal dichotomy. Singapore Journal of Tropical Geography 37(3): 282-294.

Lazar, S. and Nuijten, M., 2013. Citizenship, the self, and political agency. Critique of Anthropology 33(1): 3-7.

Nuijten, M., Bode Bakker, M. and Bok, J., 2019. Lucha y esperanza: testimonios de participantes en la PAH. Wageningen.

Nuijten, M., 2013. The perversity of the 'citizenship game': slum-upgrading in the urban periphery of Recife, Brazil.' Critique of Anthropology 33(1): 8-25.

Ross, S., Savage, L. and Watson, J., 2020. University teachers and resistance in the neoliberal university. Labor Studies Journal 45(3): 227-249.

26. MONIQUE NUIJTEN AND ETHICS OF CARE AS A ROADMAP FOR ENGAGED SCHOLARSHIP

Elisabet Dueholm Rasch

Barricades

'I don't really feel comfortable at the barricades, I'm more of the observing type,' Monique Nuijten once said to me. I thought of this as a remarkable comment for someone who could, in my view, tick all the boxes of activist anthropology. At the same time, I could also see how the comment reflected the way she gave form and content to her scholarship: She would never put herself upfront, but always make sure that the voices of others – research participants, students, colleagues – would be heard. She would do so in a way that might best be characterised as an ethics of care, the (political) act of taking care of students, colleagues, and ourselves within the context of the neoliberal university (Lawson, 2009; Mountz *et al.*, 2015). Monique not only practises such an ethics of care in her daily university life, but extends this to the way that she conducts, and writes about, her research. This element of care makes her research engaged and activist in all its facets. In this contribution, I want to celebrate this dimension of Monique's scholarship by reflecting on how she brings together ethics of care with engaged and activist anthropology and as such makes it possible to, in the words of Lawson, 'engage in radically open, democratic and transformative practices for change (Lawson, 2009: 212)'.

Solidarity

I did see Monique go to the barricades once, though. It happened when my prolonged sick leave proved not to be compatible with our university's tenure metrics. As a result, I was on the brink of losing my appointment. When Monique found out, she started organising a meeting to discuss: 'How we can support Elisabet'. She did so, because she wanted to organise as 'concerned colleagues'. In the end it all worked out with my tenure, but in the heat of the moment this act of solidarity was of immense significance to me.

Most of the time, however, Monique's ethics of care becomes manifest in her social relations with students and colleagues, in acts that might not seem exceptional at first sight. However, within a university that is increasingly focused on publication credits, metrics. acquisition for competitive grants, and individual career paths, everyday acts of taking care of others, like:

- taking the time to guide (PhD) students towards their first conference talk or publication of a paper without claiming first authorship;
- sharing ideas on how to organise research seminars as a safe space for everyone;
- formulating e-mails (being about drinking coffee together, the development of a new course, or the organisation of a panel) in a constructive, careful, and precise manner;
- making the 'SDC (Sociology of Development and Change group)-superwomen-dinner' happen;
- pushing for more feminine forms of leadership;
- organising carpooling from Utrecht to whatever place in the countryside where we would have away days as a chairgroup, and back;
- normalising talking about how one feels;
- being resolute, determined, and honest, but never pushy, nor dominant;
- making oneself vulnerable by sharing openly how one feels during and after a burn-out;

become political acts.

I do not think I ever heard Monique call herself a feminist, but the way that she navigates social relations within the walls of the university, brings to life many feminist texts about ethics of care that call for attention for *how* we work together and interact with one another (Lawson, 2009) and advocate for a feminist ethics of care to disrupt the neoliberal university (Conesa, 2018).

Being together

Monique brings these ethics of care into her way of doing fieldwork and writing about her findings. She does so by working 'from the ground', rather than starting from the perspective of official models (Nuijten, 2003: 19). This is the central pillar of her methodological approach. She combines this with three other methodological tools. The first is to look at the flow of action: to ask what is going on, why it is going on, who engages in it, with whom, when, and how often' (Wolf, 1990, in Nuijten, 2003: 55). To understand how people's consciousness engages with the world around them, she focuses on storytelling, and following Norman Long (1989) she is especially interested in the stories that are constructed in 'interface situations' – particular interactions where relations of power are expressed, contested and transformed (Nuijten, 2015).

This way of engaging with the field, has resulted in detailed descriptions of, for example, the social dynamics and power relations that are involved in land measurements in Mexico, in glimpses into the complexity of the power relations that are at play in community meetings, analyses of interactions between mortgage clients and bank officials that reveal the exercise of power. In terms of analysis, working from below for Monique also means postponing theoretical closure, and searching for other modes of interpretation and explanation which do not privilege key actors or formal systems (Nuijten, 2003). As such, she leaves space for the voices and interpretations of marginalised groups.

This (implicit) element of care in Monique's (field)work as a way of opening up avenues for thinking about power and transformative social change, resonates (with) the idea within engaged anthropology of transforming vertical power relations in and through research (Low *et al.*, 2010, Hale, 2006). 'Being together' and 'taking care' can become expressions of solidarity, and as such become central to engaged research.

It is, however, not before 2015, in her article about the PAH (Platform of Mortgage Victims) in Spain (Nuijten, 2015) that Monique explicitly identifies with engaged ethnography. Inspired by Lyon-Callo and Hyatt (2003) she moves towards 'a more interactive dialogue and collaboration (Nuijten, 2015: 481)' because she experienced that PAH activists demanded an active engagement from her in the form of a discussion partner (Nuijten, 2015, and several personal communications). Although she did not go to the barricades, she did 'find it fruitful to establish a more open dialogue with my respondents (Nuijten, 2015: 481)'. Her 'selling point' – fine-grained ethnography from below – remains central in creating new political possibilities for the people involved in her research.

Transgressing boundaries

Monique's engaged ethnography with the PAH also transgressed traditional power relations in research by rethinking the form in which she would share her findings. Next to her article in the *Journal of Legal Pluralism and Informal Law* (2015), she composed and published a booklet together with two junior co-researchers that featured poems, personal narratives, and songs that they collected with PAH members. Unfortunately, Monique has not been able not able to go back to Madrid and Malaga to hand over this booklet in person. Maritza Bode Bakker, one of the co-researchers, went in her place and has shared how the booklet was received with great enthusiasm on several occasions (also in this festschrift).

In a similar vein, we (Monique, our colleague Michiel Köhne and myself) started to experiment with different ways to cross the boundaries between our own academic work and the political actions of the groups that we work with. We did so in the guise of our Cluster for Grassroots Politics (CGP). Although the CGP in its present form is a product of many conversations, discussions, and activities that we organised together, it was Monique who was the first to see the potential of bringing us and our research together in a cluster. The connection between us was obvious content-wise, as our case studies – the PAH in Spain (Monique), *Tegengas* (a partnership against shale gas) in the Noordoostpolder, the Netherlands (Michiel and Elisabet), protests against coal mining in Australia (Michiel) and territory defenders in Guatemala (Elisabet) – all look into the dynamics of grassroot politics.

The most inspiring aspect of our work with the cluster, however, is that we have brought academics, (former) students, and activists together, in panel sessions, in a political café about resistance, and in a round-table debate about the relevance of our study programme for activism. Monique clearly loved being in the middle of these activities and would often say: 'Wow, this really gives me a lot of energy'. In the literature, such ways of producing and sharing knowledge are considered central to engaged and activist anthropology. Making knowledge accessible to the public, as well as involving research participants in the actual production of papers, presentations and lectures can be an avenue for creating more horizontal relations in the research process. In addition, it breaks through the traditional power relations of the researcher determining what to share with her or his academic audience, and how (Rasch and Van Drunen, 2017). I am not completely sure, though, whether Monique would use such phrases to explain what she does. She probably does, I think, most of these things because she cares about giving a voice to (PAH) activists and students.

No closure

The way Monique Nuijten navigates care in fieldwork and more broadly in academia shows the potential of bringing an ethics of care into our understanding of engaged anthropology. Starting from an ethics of care – taking care of others, and for yourself

– can enrich our practice of engaged and activist anthropology, and align our research practice with other elements of 'being a good academic'. It also, and I repeat Lawson here, makes it possible to engage in open, transformative practices for change (Lawson, 2009: 212).

Monique, I hope you don't mind that I now take your work and your way of enacting ethics of care in your academic life to propose a understanding of engaged, activist scholarship, in which ethics of care serves as a roadmap. Ethics of care, then, is a compass that can help explore ideas about how we can be the change we want to see inside and outside the walls of the university (Chatterton *et al.*, 2010). Michiel and I will, together with Jessica Duncan, keep this legacy of yours alive through the Grassroots Politics Cluster. No closure there, either.

Elisabet Dueholm Rasch / Associate Professor, Sociology of Development and Change Group, Wageningen University, the Netherlands / elisabet.rasch@wur.nl

References

Chatterton, P., Hodkinson, S. and Pickerill, J., 2010. Beyond scholar activism: making strategic interventions inside and outside the neoliberal university. Acme: An international e-journal for critical geographies 9(2): 245-274.

Conesa, E., 2018. How are academic lives sustained? Gender and the ethics of care in the neoliberal accelerated academy. Impact of Social Sciences Blog. Available at: https://blogs.lse.ac.uk/impactofsocialsciences/2018/03/27/how-are-academic-lives-sustained-gender-andthe-ethics-of-care-in-the-neoliberal-accelerated-academy/.

Hale, C.R., 2006. Activist research v. cultural critique: indigenous land rights and the contradictions of politically engaged anthropology. Cultural anthropology 21(1): 96-120.

Lawson, V., 2009. Instead of radical geography, how about caring geography? Antipode 41(1): 210-213.

Long, N., 1989. Encounters at the Interface. Netherlands: Agricultural University of Wageningen.

Low, S.M., Merry, S.E., 2010. Engaged anthropology: diversity and dilemmas: an introduction to supplement 2. Current anthropology 51(S2): S203-S226.

Lyon-Callo, V., and Hyatt, S.B., 2003. The neoliberal state and the depoliticization of poverty: activist anthropology and 'ethnography from below'. Urban Anthropology and Studies of Cultural Systems and World Economic Development 32(2): 175-204.

Mountz, A., Bonds, A., Mansfield, B., Loyd, J., Hyndman, J., Walton-Roberts, M., Basu, R., Whitson, R., Hawkins, R., Hamilton, T. and Curran, W., 2015. For slow scholarship: a feminist politics of resistance through collective action in the neoliberal university. ACME: An International Journal for Critical Geographies 14(4):,1235-1259.

Nuijten, M., 2015. Political activism and the PAH (Platform of Mortgage Victims) in Spain: the significance of Franz von Benda-Beckmann's work for the study of power, political agency and legal pluralism at the grassroots. The Journal of Legal Pluralism and Unofficial Law 47(3): 476-492.

Nuijten, M.C.M., 2005. Power in practice: a force field approach to power in natural resource management. Journal of Transdisciplinary Environmental Studies 4(2): 3-14.

Nuijten, M.C.M., 2003. Power, community and the state: the political anthropology of organisation in Mexico. Pluto Press.

Rasch, E.D. and Van Drunen, S., 2017. Engaged ethnography as solidarity? Exploring sites for academic-activist dialogues and collaborations in engaged ethnographic research in the field of natural resource conflicts. Etnofoor 29(2): 23-38.

27. FEELING LIKE A POLITICAL ANTHROPOLOGIST

Rivke Jaffe

It is no exaggeration to say that Monique Nuijten has been a key figure in the flourishing of political anthropology in the Netherlands over the past two decades. While arguably, we can trace a long tradition of Dutch legal and political anthropology back to the interest of colonial-era scholars in Indonesia's *adat* tradition, Monique's work has been a catalyst in the development of a more critical anthropological interest in the state, citizenship, law, corruption and political agency – in the Netherlands and far beyond.

Her ethnographic engagement with these themes was initially rooted in her work in rural Mexico on the *ejido* and struggles over land reform (Nuijten, 2003), and extended through related work on land and governance in the central highlands of Peru (e.g. Nuijten and Lorenzo, 2009) – an agrarian focus that allowed her to explore all the complexities of how state power is imagined, enacted and evaded. My own engagement with her work only really came after she made a rather significant shift to a new fieldwork site, the Brazilian city of Recife. While many ethnographers may be reluctant to change their research site to a country that requires fluency in a new language and new cultural codes, it is perhaps even more surprising to find a scholar so invested in analysing agrarian land politics refocusing her attention on urban power struggles. Yet Monique managed this transition with seemingly little effort, publishing a range of insightful work – often together with Martijn Koster – on urban space, aesthetics and political belonging.

This urban research inspires me in multiple ways that extend beyond my various collaborations with Martijn and my indirect connection to Recife through the work of Carolina Maurity Frossard, the PhD candidate he and I co-supervise.

As an anthropologist who transitioned to geography, I appreciate her sensibility to the way power works in, through and on space. Focusing on a World Bank funded 'favela upgrading' project, her work in Recife highlights how the modernist aesthetics of public housing estates are part of a regime of spatial ordering, and implicitly a project of civilisation, aimed at low-income Recifenses – who in turn may reappropriate and resignify these spaces (Nuijten *et al.*, 2012).

Related closely to this, is her careful reflection on what our surroundings and our politics *feel* like, on the affective dimension of space, power and the state. This is central to her influential work on Mexico and the state as a hope-generating machine, where she shows how desire and hope co-exist with political cynicism. But it is also clear in her work on Recife, where she shows how new housing developments may not necessarily produce the new self-regulating citizens that had been envisioned, but do foster a strong of political belonging, inclusion and care (Nuijten 2013). This urban research also shows how the inhabitants of the neighbourhoods being upgraded may experience 'informal' practices as benign and predictable, in contrast with the 'formality' of planning that is often felt to be frightening and erratic (Koster and Nuijten, 2016).

To me, this ethnographic sensibility and conceptual elaboration of political affect reflect what could be understood as a feminist methodology, an attentiveness to how power works through emotions, embodiment and aesthetics. Monique's research highlights how anthropology must analyse the political work that feeling does, and in so doing points to the centrality of feeling in our own work as researchers. I have admired the way in which she acknowledged the emotional toll fieldwork can take in contexts of extreme deprivation and violence, and her ability to develop yet another research project, this time on evictions in urban Spain (Nuijten, 2015), that again shone light on the complex amalgam of suffering, hope and action that often characterises spaces of politics. She shows us how such spaces – from common agrarian land, to an informal settlement, to a foreclosed home – move the people who live there, and in so doing, she moves her readers to reflect, to care, to act.

Rivke Jaffe / Professor of Urban Geography, Department of Human Geography, Planning and International Development Studies, University of Amsterdam, the Netherlands / r.k.jaffe@uva.nl

References

Koster, M. and Nuijten, M., 2016. Coproducing urban space: rethinking the formal/informal dichotomy. Singapore Journal of Tropical Geography 37(3): 282-294.

Nuijten, M., 2003. Power, community, and the state: the political anthropology of organization in Mexico. London: Pluto Press.

Nuijten, M., 2013. The perversity of the 'Citizenship Game': slum-upgrading in the urban periphery of Recife, Brazil. Critique of Anthropology 33(1): 8-25.

Nuijten, M., 2015. Political activism and the PAH (Platform of Mortgage Victims) in Spain: the significance of Franz von Benda-Beckmann's work for the study of power, political agency and legal pluralism at the grassroots. The Journal of Legal Pluralism and Unofficial Law 47(3): 476-492.

Nuijten, M., Koster, M. and De Vries, P., 2012. Regimes of spatial ordering in Brazil: neoliberalism, leftist populism and modernist aesthetics in slum upgrading in Recife. Singapore Journal of Tropical Geography 33(2): 157-170.

Nuijten, M. and Lorenzo, D., 2009. Ruling by record: the meaning of rights, rules and registration in an Andean Comunidad. Development and Change 40(1): 81-103.

28. LOOKING FOR HOPE IN A WORLD OF STRUGGLE

Martijn Koster

In 2000, for the first time, I walked into Monique Nuijten's office at the Leeuwenborch, the building of Wageningen University's Social Sciences. I had started preparing my thesis fieldwork on popular participation in underprivileged neighbourhoods in Cochabamba, Bolivia. Although, at that time, I did not know Monique well, she seemed to be the perfect supervisor for this project, because of her expertise in political anthropology, her interest in the relationship between marginalised populations and the state, and her knowledge of Latin America as a region. Also, fellow students had told me that she was very friendly. After talking for a while, Monique agreed on supervising my project. This, in hindsight, was the start of a long and fruitful collaboration, with me, first, in the role of a master's student, later a PhD student, and still later a colleague. Over the years, I came to know Monique as a warm person with a great sense of humour, who is genuinely interested in the people around her. I also came to know here as a professional with many strengths.

First, I would like to mention Monique's great fieldwork skills. When Monique supervised my PhD research in Recife, Brazil, I had the pleasure of doing fieldwork together. I witnessed Monique's ability to 'dig deeper'. She observed what people were doing and invited them to explain what they did, how they did it and why – combining sharp questions with a friendly atmosphere, retrieving information while putting the informant at ease (as we all learnt from our ethnography textbooks but often find hard to put into practice). Every day of fieldwork, Monique gathered enormous amounts of data. She made many notes, reflected on them and discussed her findings with me and other researchers or students. When supervising my research, she always emphasised the importance of 'getting the ethnography right'. She explained to me that the theoretical contribution of an article, a book, or a dissertation – how sophisticated and innovative it may be – will almost always become outdated with time. However, she said, the ethnographic descriptions, including people's own stories and their interpretations of the world they live in, will still be worth reading after many years. Related to this is Monique's enjoyment of working with PhDs and early career scholars. Over the years, Monique and I organised two conference panels, both on urban development. When we had to select paper abstracts, Monique usually preferred the ones that promised detailed ethnographies – resulting in panels with many young researchers and lots of enjoyable 'fresh-from-the-field' presentations.

A second professional strength is that Monique's theoretical contributions are always based on solid ethnographic data. As I said above, robust data collection is her strong suit. Moreover, through her findings, Monique advanced social theory in the fields of critical development studies and political anthropology. Her concept of the state as a hope-generating-machine (Nuijten, 2003), the force-field approach (2005), and her reflections on brokerage (2004) and citizenship (2013) have inspired me and many others. Her studies of rural (2003) and urban development (2013; Nuijten *et al.*, 2012) illuminate how structural forces work and what their consequences are for people. Often, these studies present new and unexpected views on what is going on. Indeed, Monique's ethnographies demonstrate how people's practices and perceptions challenge accepted interpretations of the world and question what is taken for granted in current structures of power and inequality.

Monique's theoretical focus, in most of her work, is on hope. In addition to being a conceptual tool, hope also guided Monique's research as a political orientation. She conducted fieldwork in difficult and seemingly desperate circumstances among marginalised and often disenfranchised people: peasants in Mexico and Peru, *favela* residents in Brazil, people who are trying to make ends meet in the face of economic deprivation and social exclusion. In these settings, Monique was always looking for glimmers of hope and creativity. Her work shows how people organise their livelihoods, their homes, and their lives. It highlights how people find ways to deal with their everyday constraints. Sometimes, this resulted in criticisms of Monique's work for being too optimistic about the potential and scope of people's practices to counter inequality and exploitation. Indeed, larger political structures, economic systems and policy arrangements may seem very deterministic, a fact that Monique never ignores. However, Monique's work also underscores the importance of showing how people's practices may challenge such structures, systems and policies. Political regimes and constellations of power are never complete in their seeming hegemony. Ethnography, more than any other research approach, has the ability to show the cracks and fissures in their constructions: it demonstrates where people's practices and imaginations defy, ignore or creatively rework structures of authority and rule. In so doing, it points us towards a space for new politics, small as it may be, a site and spur of becoming, of possibilities and hope (Gibson-Graham, 2005; Van Leerzem *et al.*, 2016). This hope also guided the research in Monique's most recent project, on activists in Spain, on which she published a booklet with the unambiguous title *Lucha y Esperanza* (Struggle and Hope) (Nuijten *et al.*, 2019). Monique is looking for hope in a world of struggle.

I am very grateful to Monique, more than I can express in this contribution to her Liber Amicorum. She helped me with my first steps as an academic. In the years that followed, she encouraged me (and many others) in my work at the university and in the field, through her intellectual guidance, her constructive critique and her warm personality. Most of all, she inspires me and many others to keep on looking for hope and have confidence in a better future.

Martijn Koster / Anthropology and Development Studies, Radboud University, Nijmegen, the Netherlands / m.koster@maw.ru.nl

References

Gibson-Graham, J.K., 2005. Building community economies: women and the politics of place. In: Harcourt, W. and Escobar, A. (eds) Women and the politics of place. Kumarian Press, Bloomfield, 130-157.

Nuijten, M., 2003. Power, community and the state: the political anthropology of organisation in Mexico. London: Pluto.

Nuijten, M., 2004. Between fear and fantasy: governmentality and the working of power in Mexico. Critique of anthropology 24(2): 209-230. https://doi.org/10.1177/0308275X04042655

Nuijten, M., 2005. Power in practice: a force field approach to natural resource management. Journal of Transdisciplinary Environmental Studies 4(2): 3-14.

Nuijten, M., 2013. The perversity of the 'citizenship game': slum-upgrading in the urban periphery of Recife, Brazil. Critique of Anthropology 33(1): 8-25. https://doi.org/10.1177/0308275X12466683

Nuijten, M., Koster, M. and De Vries, P., 2012. Regimes of spatial ordering in Brazil: neoliberalism, leftist populism and modernist aesthetics in slum upgrading in Recife. Singapore Journal of Tropical Geography 33(2): 157-170. https://doi.org/10.1111/j.1467-9493.2012.00456.x

Nuijten, M., Bode Bakker, M. and Bok, J. (eds), 2019. Lucha y esperanza: testimonios de participantes en la PAH. Wageningen, Wageningen University & Research.

Van Leerzem, L., Nuijten, M. and De Vries, P., 2016. Being *negro* in Recife Brazil: a political aesthetics approach. Ethnos 81(5): 865-887. https://doi.org/10.1080/00141844.2015.1028565

29. ON STORYTELLING IN EDUCATION – MONIQUE AS INSPIRATION

Jessica de Koning

Upon walking into the classroom, Monique Nuijten puts down her bag, organises her printed-out lecture slides, and starts the PowerPoint.. Her slides do not start with a clear structure of today's content. Monique does not walk through the classroom. She does not use different types of teaching tools to address the various students. I cannot identify energising exercises during the lecture. Many of the elements, tools, and tricks Wageningen University teaches us to enhance our teaching skills seem to be missing. Even though she is soft-spoken, students listen attentively to Monique's personal stories about experiences in the field. When I watch her teach, I realise that it is these stories that contribute to the learning experience of the students. By telling stories, Monique created an atmosphere where students felt safe and more willing to take part.

Storytelling in higher education is increasingly accepted as a valid way to teach (Molthan-Hill *et al.*, 2020). Storytelling enables teachers to transcend information and enables students to make sense of experience, stimulate our imagination, and enhance our memory (Alterio and McDrury, 2003; Wall *et al.*, 2019). In today's higher education, captivating students' attention and ensuring students remember the study materials for a longer period is one of the more challenging tasks. Storytelling is a powerful tool because stories stick. Stories allow for factual, visual, and emotional processing of information (Wall *et al.*, 2019; Molthan-Hill *et al.*, 2020). Already in 1969, Bower and

Clark (1969) found that students learning through stories can recall facts six to seven times more than students who do not use stories.

Moreover, storytelling stimulates critical thinking, self-reflection, and contributes to conveying practical realities (Abrahamson, 1998; Atlterio and McDrury, 2003, Molthan-Hill *et al.*, 2020). Critical thinking is particularly relevant in a chair group such as Development and Change with a scientific interest in inequality, marginalisation, and political agency. While stories about experiences might not offer a tangible approach to solving inequality, they can empower learners to invoke new ideas through their imagination and to construct new visions for the future (Molthan-Hill *et al.*, 2020; Wall and Perrin, 2015; Wall, 2016).

Nowadays, education at Wageningen University is increasingly digital because of shifts in education strategies, development of digital education tools, and a global pandemic. This creates a few challenges. The distance between teachers and learners is growing as physical and personal interactions are less frequent. Teaching is more instrumental; focusing increasingly on transparency and accountability. Teachers struggle with connecting to students and creating an enabling safe space for learning. To remedy this, Wageningen University offers more courses on student engagement, intrinsic motivation, and brain-based teaching (https://esd.crs.wur.nl). Within these developments, storytelling has the potential of becoming a vital element as it addresses the current challenges of teaching. Storytelling increases the social connection students make with teachers, it creates a safe space for learners and diminishes the distance between teachers and learners. Even when the space and place of teaching change from on campus to online, storytelling sticks and works.

Monique already understood this years ago as she told her stories, captivated the students, and inspired me to do the same.

Jessica de Koning / Education Coordinator, Rural Sociology & Sociology of Development and Change, Wageningen University, the Netherlands / jessica.dekoning@wur.nl

References

Abrahamson, C.E., 1998. Storytelling as a pedagogical tool in higher education. Education 118(3): 440-452.

Alterio, M.and McDrury, J., 2003. Learning through storytelling in higher education: using reflection and experience to improve learning. Routledge.

Bower, G.H. and Clark, M.C., 1969. Narrative stories as mediators for serial learning. Psychonomic Science 14(4): 181-182.

McLellan, H., 2007. Digital storytelling in higher education. Journal of Computing in Higher Education 19(1): 65-79.

Molthan-Hill, P., Luna, H., Wall, T., Puntha, H. and Baden, D. (eds), 2020. Storytelling for sustainability in higher education: an educator's handbook. Routledge.
Wall, T. and Perrin, D., 2015. Slavoj Žižek: A Žižekian gaze at education. London: Springer.
Wall, T., Clough, D., Österlind, E. and Hindley, A., 2019. Conjuring a 'spirit' for sustainability: a review of the socio-materialist effects of provocative pedagogies. In: Leal Filho, W. and Consorte McCrea, A. (eds) Sustainability and the humanities. Cham: Springer.

PART 3:

PERSONAL ENCOUNTERS

In which eleven colleagues draw on their personal encounters with, and recollections of, Monique Nuijten painting a picture of her as an inspirational person, whose embodied scholarship is an integrated part of her life. As in part 1 and 2 of this book, it becomes evident how her personality and scholarship intersect to make a lasting mark on people and knowledge.

30. TAKING PRACTICE SERIOUSLY

Gemma van der Haar

Meeting Monique

I somehow knew Monique Nuijten before ever having met her. When I joined the Rural Development Sociology (RDS) group at Wageningen University as a starting PhD candidate, Monique and her family were in Mexico, but the references made to her were many and struck me for the warm appreciation they conveyed. Over the years, these expressions of appreciation, admiration and respect were a constant. I think it was Raymond Buve, professor of Mexican history in Leiden, who told me, referring to Monique: 'Now here you have someone who really has something to say'.[10] Even then, at the time of concluding her PhD, she seemed to me such a mature scholar: always professional, always articulate, and always a step ahead of everyone else in her thinking.

During the write-up phase of my PhD, I drew a lot of encouragement from my interactions with Monique. Being back at the office after a long fieldwork period, in

[10] His words were, in Dutch: 'Dit is iemand, die kan écht iets'.

which I had been largely disconnected from what happened in Wageningen, I did not immediately feel at home in the RDS environment. RDS seminars could get rough, but even then I found what I heard of Monique's work very inspiring and her feedback on my own presentations was invariably rich. She and Pieter, between the two of them, always seemed to have read the most exciting new books and papers and often suggested new avenues for analysis.

Debating the state

Like Monique, my PhD was on Mexico, and like hers, it was on land. The two settings we researched were very different, however. The outmigration to the US that was so marked in her case (in Jalisco), for example, was at the time almost entirely absent from where I worked in Chiapas. Though like Monique I studied *ejidos* (landholding communities created through land redistribution), in my case these were indigenous (Maya) and seemed, at times, to present a rather different universe. Some of our most interesting conversations related to the very different experiences people seemed to have of the state. In her PhD thesis (Nuijten, 1998) Monique focused on how people sought to connect to the state, and how the state was both very present and very elusive at the same time. I came out of my fieldwork stating that for all practical purposes, in daily life at the village level, 'there was no state'. Resisting at first Monique's insistence on the state, I gradually started to be sensitive to the many ways in which the state, even if it had no active presence in everyday life, was a reference, was part of people's horizons, was very present, as she put it (citing Abrams, 1988 [1977]), as an *idea*. This allowed me to develop my analysis of the multiple presences and absences of the state in local notions of property, community, and political agency.

A liberating legacy: organising practices

What has stood out most for me personally in Monique's work is the notion of 'organising practices', defined by Monique herself in *Power, community and the state* (2003: 11) as the 'manifold forms of organising, individual or collective'. From the first time I heard her take on this, it immediately resonated. I recognised it as liberating. To me, the notion created the room to take people's practices seriously in all their complexity, not as 'derivatives' of structure or as mere epiphenomena of something 'more real' lurking underneath. The notion of organising practices allowed me to make a conceptual leap from 'organisation' to 'organising', to rescue organisation from the confines of structure, form and norm, and to study it as process, as always emergent, as society always in the making. This was about people doing and undoing: making and unmaking order, finding and abandoning form, producing and disrupting alliances, weaving and unravelling webs of meaning. The concept as Monique coined it, defies reductionism. It takes 'organisation' out of the normative realm, where people are being assessed as either 'well-organised' or not. It takes people's efforts to try and make sense of the world seriously, embracing rather than explaining away the contradictions and ambiguities in such sensemaking.

In my understanding, for Monique the focus on organising practices was both a theoretical and an empirical proposal: through the study of practices we could gain an understanding of the workings of power. I continue to find this a very productive standpoint. I have found myself re-working this focus on practices in different research projects. First of all, by proposing to study 'governing practices' in my postdoc on contested governance in the context of the Zapatista uprising in Chiapas, where I wanted to look at the 'doing' of governance to analyse competing claims to legitimacy and authority (Nuijten *et al.*, 2004). More recently I have proposed, with colleagues Lotje de Vries and Mathijs van Leeuwen, to study land access and land conflict in terms of 'claim making practices' (Van der Haar *et al.*, 2020). It was not until now, preparing this text, that I realised that this is yet another instantiation of the same conceptual move, this time from 'claims' to 'claim *making*', and with a similar non-reductionist approach to grassroots political agency.

Monique's work drove home for me the power of ethnography and made it possible to legitimise in hindsight some of the choices I had almost intuitively made in my own fieldwork. What to me makes Monique's work so powerful is the combination of ethnographic sensitivity and her capacity and audacity to engage with big questions. She has had, I found, a healthy irreverence to big names and big concepts whenever she did not find them convincing. This, too, I found very liberating.

Navigating the work floor together

If you walk the hallway of 'our' part of the third floor in the building that houses the social sciences at WUR (the Leeuwenborch) today, you would come across a poster in a bold red-and-black design saying 'The urban turn'. It is the lasting memory to an event Monique and I organised a few years back, together with Bram Jansen and other SDC colleagues, in response to a growing interest and in converging research agendas on the urban within SDC (full title: The urban turn in development sociology: Ordering and agency in the city). It is in every way fair to say that Monique had the lead in the event: she had the most well-articulated agenda, brought in some of the main speakers, and the poster I mentioned was designed based on her suggestions, and it was an absolute pleasure to collaborate on it with her.

This event marks for me also the one advantage of what was otherwise more than anything a painful and damaging process, that of the (involuntary) merger of the unit at which I worked (the Disaster Studies group) with the bigger Rural Development Sociology group where Monique was based.[11] After my PhD, I had been out of Wageningen University for a while and when I returned it was to join the unit of Disaster Studies which had branched off from RDS years before. One good thing of the merger was that Monique now became a direct colleague and that there were more opportunities to collaborate than before. Monique felt like an ally to me. We often shared a similar analysis of things happening, in the group as in the world, and our discussions were inspiring and encouraging. It was also Monique who insisted that the female staff members at SDC should pull together more. Not long before she fell ill, it was on her initiative that several of us had a wonderful dinner at the Italian restaurant in Bennekom.

No closure

One of the very useful things Monique told me once after a seminar is this: 'Don't look for closure. It is about opening up, suggesting new questions'. But at least once she did not keep to her own advice. The email she wrote to tell her colleagues and students that she had received bad news on her health, included a very definitive goodbye. Fortunately, the closure turned out not to be as absolute as it seemed at the time and we have had the privilege of Monique's company for a while longer. This gave me the time to tell her how much I missed her presence at the group.

I want to end by celebrating Monique's legacy. There is no closure there. Just as the interaction with Monique opened up new avenues for me, I am sure she has also inspired many other colleagues and many of her students.

Gemma van der Haar / Sociology of Development and Change, Wageningen University, the Netherlands / gemma.vanderhaar@wur.nl

References

Abrams, P., 1977. Notes on the difficulty of studying the state. Journal of Historical Sociology 1(1): 58-89.

Nuijten, M., 1998. In the name of the land: organization, transnationalism, and the culture of the state in a Mexican ejido. PhD Dissertation, Wageningen University.

Nuijten, M., 2003. Power, community and the state: the political anthropology of organisation in Mexico. London: Pluto.

Nuijten, M., Anders, G., Van Gastel, J., Van der Haar, G., Van Nijnatten, C. and Warner, J., 2004. Governance in action: some theoretical and practical reflections on a key concept. In: Kalb, D., Pansters, W. and Siebers, H. (eds) Globalisation and development. Springer: Dordrecht.

Van der Haar, G., Van Leeuwen, M. and De Vries, L., 2020. Claim-making as social practice: land, politics and conflict in Africa. Geoforum 109: 111-114.

[11] This merger produced the group now known as SDC: Sociology of Development and Change.

31. THE ETHNOGRAPHY OF POWER AND THE POWER OF ETHNOGRAPHY

Jilles van Gastel

Early 2000, after an internship at an INGO in Vietnam where I was confronted with the gap between policy and implementation, I set my mind on studying the policy-making process in an aid organisation for my master thesis. My initial supervisor suggested getting a second advisor with expertise on policy and organisation processes and mentioned Monique Nuijten´s name, a postdoc in the group of Rural Development Sociology. I hadn´t heard of her nor did any of my fellow students as she didn´t have any teaching obligations, nor had she supervised any students yet. Not knowing how this would work out, I decided to take the chance and ask her if she was interested in helping guide my thesis. That was the start of a long-lasting mentorship and friendship.

Although I had been trained to work and do research abroad, I decided I wanted to understand how international aid organisations make development policies. Fortunately, I had the opportunity to do fieldwork at the Dutch Ministry of Foreign Affairs in the area of development cooperation. With her consultancy experience at the FAO, Monique had briefly tasted that world and because of her studies in development sociology, she understood that these organisations are yet another world with their own jargon, knowledge, logic, and rules. During the first weeks of that field work in the Ministry, I often felt lost, not knowing what to look at and focus on as policy and policy-making seemed to be everywhere and nowhere at the same time. Monique then taught me one of the most important lessons in my academic career: where you find that things do not make sense, what puzzles, frustrates or surprises you: *that* is where to focus the research on and where the researcher can gain new insights. Now it seems so evident, but for a young and unexperienced student starting her first fieldwork, this simple advice turned out to be an excellent guide for my research; it taught me to use my body as an instrument in fieldwork, and emotions, whether pleasant or not, as my internal compass for research.

Monique guided her students with dedication, patience, and openness. She read each text I sent to her and provided comments that aimed at helping organise my ideas and add more structure to my arguments. Regarding theoretical orientation, Monique was careful not to impose her own conceptual framework and instead suggested reading books of authors that inspired her and that she hoped would stimulate my thought process. The drive behind her work is the question of how larger processes and practices of bureaucracies form and shape the human experience. The literature that I believe inspired her work most were well-written ethnographies about the everyday lives of people. Whether it is about the *ejidatarios* in Mexico, *favelados* in Brazil, *campesinos* in the highlands of Bolivia, or *pastores* in Spain, the detailed description of their daily routines and struggles in the wider socio-political and economic context gives us a glimpse of how people make sense of their lives and give meaning to it.

Ethnography provides us with a compelling narrative of what it is to be human and stimulates our empathy for the other, particularly but not exclusively the marginalised and vulnerable. However, good ethnography goes beyond that and provides new insights into theoretical issues. Monique does that forcefully; she has shown how the Mexican bureaucracy deals with land conflicts between peasants and private landholders, and, through techniques of governmentality, produces and reproduces fears and fantasies about possible resolutions of the conflict (Nuijten, 2004). Monique captures these practices convincingly with the concept 'hope-generating machine' (Nuijten, 2003).

After my master thesis *Governing 'Good Governance'*, a case study on policy making in the Dutch ministry of Foreign Affairs, Monique invited me, together with others, to work on a chapter in the book *Globalization and Development* (Kalb *et al.*, 2004). Monique gave space to early-career scholars and united them in their shared concern for critical thinking on development and interest in issues of order, rule, and power, while respecting their diversity in research topics and disciplinary focus. In line with critical development studies, we unpacked the governance concept. Instead of defining *the* governance concept and developing a single theoretical framework, we show how we each deal with the concept in our individual research projects and where we see commonalities in our analyses; in the rise of the concept in policy and academic circles as a result of neo-liberal reforms, privatisation and globalisation, in the often concealing nature of the concept and the importance to address power relations, and in the dilemma of engaging in the dominant discourse to change it or developing our own analytical categories. One of the key reasons why this chapter ´Governance in Action´ (Nuijten *et al.*, 2004) achieved this collective reflection is because after each working session, we shared a drink and some finger food only to continue our conversations on a more personal level, talking about our experiences in research and academic life. Monique´s efforts to bring people together and create affectionate relations made this collaboration a success.

In that two-year period between my Master thesis and PhD research, Monique generously introduced me to the European Inter-University Development Opportunities Study-Group (EIDOS), a group of British, Dutch, and German academics interested in the critical study of development. At its conference in London, we presented a paper based on my thesis that became a chapter in the book edited by David Mosse and David Lewis called *The Aid Effect: Ethnographies of Development Practice and Neo-liberal Reform* (Mosseand Lewis, 2005). Then she supported me by helping me to draft a proposal to apply for a scholarship of the Netherlands Organization for Scientific Research (NWO) after which I became her first PhD student. Over the next years, we developed a close working relation and friendship where we connected in our quest for comprehending how others – people we met in the field but also in our everyday lives – understand, make sense of, and act upon critical events in their lives, their relationships, and their selves. I believe that for both of us this was a way to figure out our own lives and relationships. I cherish those talks with Monique that were often accompanied by a home-cooked meal as we both enjoy good food and cooking. When we were in Washington D.C. for the American Anthropological Association (AAA) conference, she took the group to a Peruvian restaurant to introduce us to a tasty *ceviche* and *lomo salteado*, my first encounter with this Latin American cuisine. At such conferences, Monique made sure everyone participating in her sessions felt part of the group. Later, because of health reasons, she needed to balance her concern for others with taking care of herself.

While Monique went through changes in her personal life, I also embarked on a new journey in 2010. Sharing my time between Chile, my new home, and the Netherlands while still trying to finish my PhD wasn´t easy. Being far apart and each working out our new lives, our talks were shorter and less frequent. When I finally finished my PhD and came to the Netherlands to defend my thesis Monique couldn´t be there. I understood but felt sad for not being able to celebrate this together and to thank her then and there for the continuous support I received, even when she was having a difficult time. After a number of years in academia in Chile, I left to go back to the sphere of development practice. Yet, Monique forever changed my thinking about development bureaucracy and its workings as a ´hope generating machine´, not as a cynical analysis or pessimistic outlook, but actually as a call to finally deliver on its promise.

Jilles van Gastel / former PhD student, Sociology of Development and Change Group, Wageningen University, the Netherlands / WorldWide Fund for Nature (WWF), Chile / jilles.vangastel@wwf.cl

References

Kalb, D., Pansters, W. and Siebers, H. (eds), 2004. Globalization and development. Themes and Concepts in Current Research. Dordrecht: Kluwer Academic Publishers.

Mosse, D. and Lewis, D. (eds), 2005. The aid effect: giving and governing in international development. London: Pluto Press.

Nuijten, M., 2003. Power, community and the state: the political anthropology of organisation in Mexico. London: Pluto.

Nuijten, M., Anders, G., Van Gastel, J., Van der Haar, G., Van Nijnatten, C. and Warner, J., 2004. Governance in action: some theoretical and practical reflections on a key concept. In: Kalb, D., Pantsers, W. and Siebers, H. (eds) Globalization & development: themes and concepts in current research. Dordrecht Kluwer Academic Publishers, pp. 103-130.

Nuijten, M., 2004. Between fear and fantasy: governmentality and the working of power in Mexico. Critique of Anthropology 24(2): 209-230.

32. VROUWEN IN DE WETENSCHAP

Nuray Tümer

Mijn meest duidelijke beeld van Monique Nuijten is die van een vrouw die staat, loopt, handelt en straalt binnen een masculien systeem, waar ze zich weinig van aantrekt. Ze had de handels- en werkwijze van dat systeem al geanalyseerd en gedeconstrueerd tot voor haar werkbare proporties als vrouw en professional, toen ik als PhD student aan de vakgroep startte met mijn onderzoek over hoe kinderen van migranten, in Nederland geboren en getogen, hun eigen thuis en zelf construeerden binnen de weinig uitnodigende omgevingen waarin ze verkeerden.

Het meest wezenlijke wat ik van Monique heb geleerd is hoe systemen te deconstrueren om vervolgens daarin je eigen positie duidelijk te krijgen, in te nemen en van daaruit met het systeem om te gaan. Een paar van Monique's ideeën die mij het meest zijn bijgebleven en een basis hebben gegeven zijn de volgende thema's, weergegeven zoals ik het mij herinner en het door de jaren heen onthouden heb.

Positieve discriminatie van vrouwen in de wetenschap – in de woorden van Monique – 'is nu eenmaal nodig'. Dit nodigt uit tot het loskoppelen van je eigen kwaliteiten van die van de je omringende systemen, die als beperkend kunnen werken. Het nodigt ook meteen uit om de positieve en verstevigende krachten van je omgeving te erkennen en daar gebruik van te maken.

Over moederschap op de werkvloer kan ik mij de woorden 'moet nu eindelijk maar eens meegenomen worden als een realiteit' herinneren. In haar curriculum vitae heeft Monique als enige vrouw die ik ken, haar 'gat' in haar CV gedicht door de geboorte en het opvoeden van haar dochters te vermelden. Hiermee, wat mij betreft, heeft ze het opvoeden van kinderen in hun meest afhankelijke jaren, als volledig onderdeel van een carrière neergezet in plaats van een in geheime kamers uitgevoerde clandestiene operatie.

Monique, dankjewel voor je eigen kijk en wat je met je meebracht en mij getoond hebt: hoe het ook kan zijn op een werkvloer als vrouwelijke professional. Ruimte voor passie, sociale omgang en het leven zelf. En dat output hiermee in positieve verhouding dient te staan om een zeer gespannen en ongezonde relatie te voorkomen.

Nuray Tümer / former PhD student, Sociology of Development of Change Group, Wageningen University, the Netherlands / Owner at the Zaaier, Wageningen, the Netherlands / nuraytumer@yahoo.com

33. FINDING YOUR WAY AS A YOUNG ADOLESCENT AFTER GRADUATION – AND REFLECTIONS ON THE ROLE MY TEACHER/SUPPORTER/FRIEND MONIQUE NUIJTEN PLAYED IN MY SEARCH

Lynn van Leerzem

When we are in our early twenties, I think we all need some people who, sometimes for unclear reasons, just seem to believe in you. Monique has played such a role in that phase of my life. No matter what I did, she seemed to think it was fascinating. In all the steps I made after graduation, she encouraged me, and she made it feel safe for me to try new things and do what fit me.

After working for a year and a half at the Ministry of Foreign Affairs, I decided to devote my time to grassroots initiatives around climate, the energy transition and community building. Monique and I had always stayed in touch after my graduation, initially to publish an article together about my thesis research in Recife, together with Pieter de Vries. But also just to chat about our lives. Monique was very interested in all the transition projects some colleagues of mine and I set up in The Hague. She would visit some of our events and interview us. She inspired us with examples of similar grassroots initiatives in Spain and elsewhere.

One moment in that period was particularly decisive for me. We were sitting at her office, and we talked about a PhD opportunity that seemed to have my name written on it: It matched my Master's thesis topic in an almost identical way (it was on the use of culture, resistance and black empowerment in Brazilian favelas). She already held the phone in her hand to call the PhD supervisor to recommend me. Meanwhile, I was explaining that I doubted to apply, that in my heart I felt I had to take the more applied, community-building path. She put the phone down and said that it was clear: I had to follow that intuition.

For many young people, I think, the phase that comes after graduation can be quite overwhelming. It is a time in which you have to make what feels like big life choices, after years of following fascinating courses and taking steps that were mostly laid out for you by others. I hope Monique knows how encouraging her support in this phase has been for me, and for many other students as well, I am sure.

There is another, more rocky way in which I have learnt a lot from Monique. This has to do with the burnout she had in that period, and particularly with her openness in describing her recovery and the struggle she had with balancing her energy and taking good care of herself. Every time she talked about this, I realised that she and I are quite alike. We both care about injustices and people around us. We see the world in an extremely detailed way. We are both perfectionists. And at times we are perhaps more interested in what is going on with other people than in taking care of ourselves. I had already outrun myself various times at that age (around 25). Hearing her stories and her warnings, I realised that if I would not start taking care of myself, it was inevitable I would also end up in a burnout or something similar sooner or later.

Speaking of Monique's care for people: there is a community somewhere on this planet, named Chão de Estrellas (Recife), where I suspect Monique will always be loved. When I lived there, people would often talk about this *gringa* from the Netherlands. They were very fond of her, appreciating her friendship and the time she took to really understand their realities. During my thesis research in Recife, Monique also pointed out to me how I was regularly ignoring my own health and wellbeing. These were simple things, like asking my host family for a better pillow as I had not been sleeping well for my first months. Or deciding not to eat with the host family 3 times a day (which meant eating dry rice with beans three times a day). This seems simple, but to consider my health as important as my respect for my host family, somehow had not occurred to me until that moment. She pushed me to stand up for myself in ways that I had not even considered as options for myself.

I wish for all young, perhaps somewhat insecure, women to have a role model such as Monique who shows you how you can be both kind, successful and incredibly strong as a woman. And that perfectionism and empathy can be very powerful strengths, as long as you also show some love for yourself. No matter how many inspiring publications you write, the mark you leave behind as an academic might just be in your soft skills and in the example you set for the young people you work with.

Lynn van Leerzem / former MSc student, Sociology of Development and Change Group, Wageningen University, the Netherlands / Milieudefensie, the Netherlands / lynn.van.leerzem@milieudefensie.nl

34. ONZE LEUKE EN LEERZAME SAMENWERKING

Maritza Bode Bakker

Ik wil sowieso als eerste zeggen: dankjewel Monique! Tijdens mijn Master Internationale Ontwikkelingsstudies, en ook daarna, heb jij mij veel geholpen en heb ik hele interessante, mooie dingen meegemaakt dankzij jou. Ik ben daardoor veel ervaringen en vriendschappen rijker, en daar ben ik je heel erg dankbaar voor.

Ik leerde je in eerste instantie kennen als docent bij een aantal vakken. Ik wist toen niet hoeveel je voor mij zou gaan betekenen. Eén vak zal mij altijd bijblijven omdat het dé inspiratiebron werd voor mijn thesis. Dat was ook echt het begin van onze samenwerking. Ik vond de documentaire 'Rize' erg interessant die je tijdens de cursus liet zien, en ging daarom daarna met jou kletsen over mijn idee om voor mijn thesis iets te doen met breakdance. Je hebt me toen een paar tips gegeven en aangegeven wie misschien mijn thesis zou kunnen begeleiden. Eenmaal in Ecuador om veldwerk te doen voor mijn afstudeerproject, moest ik opeens op zoek naar een nieuwe begeleider. Ik was ontzettend blij om jou toen te spreken, en vooral ook om te horen dat jij graag mijn onderzoek verder wilde begeleiden. Wat toen begon als begeleiding bij mijn afstudeerproject, groeide later uit tot meer, en daar ben ik heel blij om.

Tijdens mijn tijd in Ecuador hadden we vooral contact via de mail en af en toe via Skype. Je gaf me goed advies, waardoor ik uiteindelijk heel erg blij was met mijn thesis resultaat. Ik vertel nu nog graag over mijn thesis aan familie en vrienden. Ik vond het ontzettend leuk om een aantal maanden een breakdance crew te volgen en te leren over hun leven. Tijdens dit proces zag ik wat dans kon betekenen voor jongeren. Dans is een deel van hun identiteit. Daarnaast was een belangrijk thema hoe de groep door middel van dans een leven van protest leidde: protest tegen het 'normale' leven, de standaarden van de 'gewone samenleving', gendernormen en het imago van 'de straat'. Het dansen is voor hen een nieuwe levensstijl.

Toen ik terug in Nederland was, had ik een gevoel dat ik meer moest doen met al het interessante materiaal dat ik had verzameld. Ik wilde naast mijn thesis nog iets schrijven om te laten zien dat breakdance niet iets negatiefs is. Ik kwam bij jou met dit idee en je was gelijk erg enthousiast. Toen zijn we echt meer gaan samenwerken en hebben we een heel mooi artikel geschreven. Tijdens het schrijven van dit artikel werkten we eigenlijk perfect samen: Jij schreef het meer theoretische deel, en ik bracht de verhalen van de breakdancers in. Dit werkte echt heel goed. Zo kwamen we uit op een analyse van breakdance als een subcultuur van verzet en de rol van het lichaam hierin. We krijgen zelfs nu nog mails dat ons artikel goed wordt beoordeeld. Dat vind ik ontzettend gaaf! Dit is voor jou één artikel van een hele hoop waarschijnlijk, maar voor mij was het echt een hele leuke en unieke ervaring. Dat heb ik zeker jou aan te danken. Ik ben heel trots op ons werk en ik heb dit voorbeeld zelfs bij het sollicitatiegesprek voor mijn huidige baan gebruikt!

Ik was al ontzettend dankbaar voor al je hulp tijdens mijn thesis, en toen kwam je met nog een idee om onze samenwerking verder voort te zetten. Toen ik klaar was met mijn studie heb ik lang gezocht naar een baan. Op een gegeven moment zocht je contact met mij op met de vraag of ik jou zou kunnen helpen met je onderzoek in Spanje. Ik vond het heel leuk dat je aan mij dacht voor deze opdracht, en het was een hele gave een leerzame ervaring. Ik heb voor het onderzoek in Spanje een paar maanden in Madrid doorgebracht en net als in Ecuador heb ik daar ook hele bijzondere vriendschappen opgebouwd. Onze samenwerking was ook in dit project goed, we hielden dezelfde verdeling van theorie en praktijk aan. In Madrid heb ik onderzoek gedaan naar een organisatie genaamd de PAH *(Plataforma de Afectados por la Hipoteca*/Platform voor degenen die getroffen zijn door de hypotheek). Dit was een uitbreiding van jouw onderzoek naar dezelfde organisatie in Malaga en Salobreña. Ik heb met heel veel mensen gepraat over hoe zij in de problemen zijn gekomen door het niet meer kunnen betalen van hun hypotheek en hoe de PAH hun hierin heeft geholpen. Ik focuste daarbij op de rol van de PAH voor (met name Ecuadoriaanse) migranten.

Over dit onderzoek hebben we geen wetenschappelijk artikel geschreven, maar iets heel anders waar onze onderzoeksparticipanten en ikzelf ontzettend blij mee zijn! Samen met Jeroen, die het veldwerk deed in Valencia, maakten we een boekje over de PAH: '*Lucha y Esperanza: Testimonios de participantes en la PAH*' (Strijd en Hoop: Getuigenissen van PAH-participanten). We maakten dit boekje om iets terug te kunnen geven aan iedereen die zijn/haar verhaal met ons had gedeeld. We gebruikten het veldwerkmateriaal dus niet om iets theoretisch te schrijven, maar juist om de verhalen van mensen uit te lichten en tot leven te brengen. Dit boekje staat, samen met het breakdance artikel, in mijn boekenkast en daar ben ik heel trots op.

Een paar jaar geleden ben ik teruggegaan naar Madrid om het boekje te presenteren. Uiteindelijk heb ik ook het boekje aan de onderzoeksparticipanten in Malaga (waar jij zelf veldwerk had gedaan) kunnen geven toen ik daar op vakantie was. Dat was een ontzettend mooi moment, ik zag toen wat voor indruk je had achtergelaten op de mensen van de PAH in Malaga. Je hebt dus niet alleen veel invloed gehad op mij, in een periode waarin ik begon met mijn werkende leven. Je bent ook heel belangrijk voor de mensen waarmee je onderzoek doet.

Als ik zo terugkijk, komen er veel herinneringen boven, die ik heb kunnen opbouwen dankzij jou. En daarom wil ik je dus heel erg bedanken. Je had veel vertrouwen in mij en mijn toekomst. Ik heb lang gezocht naar een baan en vaak vroeg je me hoe het daarmee ging. Je hebt ook een paar keer aangegeven dat als ik een PhD zou willen doen, jij me daar ook zeker in zou willen begeleiden en dat je dacht dat ik dit heel goed zou kunnen. Dat was natuurlijk heel leuk om te horen – ook al heb ik uiteindelijk een ander pad gekozen. Ik wist niet zeker of het theoretische deel van een PhD onderzoek echt iets voor mij was. Daarom vond ik het juist heel leuk en fijn om samen te werken aan het artikel over breakdance en het onderzoek in Madrid; we vulden elkaar heel goed aan.

Waarschijnlijk heb je het al door: ik ben erg blij dat we zoveel hebben kunnen samenwerken. Ik weet dat de laatste jaren erg zwaar zijn geweest voor jou, maar ook in die tijd ben je heel belangrijk geweest voor mij, en zeker ook voor veel andere mensen. Dus … bedankt.

Maritza Bode Bakker / former MSc student, Sociology of Development and Change Group, Wageningen University, the Netherlands / TOPdesk, the Netherlands / m.bodebakker@gmail.com

35. MONIQUE NUIJTEN – UMA ORIENTADORA A SER SEGUIDA

Augusto Antonio Campelo Cabral and Mônica Valéria dos Santos Cabral

Aproximadamente, em meados de 2003, em um sábado ou domingo, não lembro bem, chega em nossa residência seu Ovídio de Paula, um senhor hoje com 84 anos de idade, líder comunitário, acompanhado de um jovem de barba avermelhada e feições europeias. Era um estudante de doutorado – Martijn Koster, de origem holandesa, que estava iniciando sua pesquisa de campo sobre as lideranças políticas (representantes comunitários) da Comunidade Chão de Estrelas, no Recife-PE, Brasil, onde naquela ocasião ele nos foi apresentado. Naquele momento eu (Augusto) era estudante do Curso de Mestrado em Geografia da Universidade Federal de Pernambuco (UFPE).

Ao longo dos meses foi surgindo uma amizade, passando a ser frequente nossos encontros – os dois estudantes para conversar, pois o objeto de pesquisa era o mesmo (a Comunidade Chão de Estrelas). Em uma destas visitas, Martijn traz seus orientadores, a professora Dr.ª Monique Nuijten e o professor Dr Pieter de Vries. A prof.ª Monique veio a ser desde então uma grande amiga, além de minha coorientadora no doutorado. Pieter um amigo que sempre quando vem ao Recife nos visita, em outrora, estávamos conversando e falando a respeito de minhas pesquisas, além da que desenvolvia no mestrado. A professora Monique mostrou-se interessada em conhecer o meu orientador, o professor Dr Jan Bitoun, que em momento posterior iríamos nos encontrar na universidade.

Encontro que ocorreu no início do ano, aproximadamente entre os meses de março ou abril de 2008, já que as provas de seleção seriam realizadas no mês de dezembro do corrente ano. Neste encontro surgiu a possibilidade de caso eu fosse aprovado na seleção para o doutorado em geografia, os professores fariam uma parceria entre o Departamento de Ciência Geográfica da UFPE com o Sociologia do Desenvolvimento, Wageningen Universidade, Holanda, e que ambos seriam meus orientadores, fato que se concretizou na seleção de 2008, com minha aprovação, iniciando as aulas no mês de março do ano seguinte (2009). É a partir desse momento que o meu contato se intensifica com prof.ª Monique, que agora, além de amiga, também é coorientadora.

A professora Monique ao passo que dava orientação, também desenvolvia suas pesquisas de campo na referida comunidade, contando com o auxílio de Mônica, sua amiga e minha esposa, que a orientava. Mônica por ser moradora do bairro de Campina do Barreto desde seu nascimento, conhecia a comunidade pesquisada, sobretudo a maioria dos moradores, onde passou a ser o braço direito para com os 'informantes-chave'. Quando Monique ia a Holanda, quem ficava responsável por ajudá-la a colher informações a respeito da pesquisa de campo, além de contatar os informantes, era Mônica.

Essa relação criou um laço muito forte de amizade entre elas, de forma que a professora Monique quando estava no Recife, fazia questão de passar dias junto com Mônica na pesquisa de campo, ou em sua casa – momento este em que Monique degustava de saborosos pratos nordestino feitos por sua amiga. Outros momentos eram dedicados à conversas com seu orientando a respeito da pesquisa de doutorado. Quando Monique não podia estar no Recife, a orientação era via e-mail.

Desta maneira, a convivência foi um ciclo de prazer e aprendizagem, pois suas observações sempre foram bastante pertinentes e perspicazes, carregada de detalhes e com questionamentos, que sem cessar, nos deixavam alertos às reflexões. Tudo isto sem perder a ternura, o sorriso, a atenção. Ela deixou de ser apenas uma orientadora, passou a ser uma amiga, uma pessoa da nossa família. Ela sempre esteve preocupada em superar as dúvidas e aflições de estudantes-pesquisadores. Por isso, queremos concluir com este pequeno e singelo texto, que todas estas palavras declaram um sentimento de profunda gratidão e alegria pelo privilégio de termos lhe conhecido, mais que isso, de ter sido sua auxiliar (pelas palavras de Mônica) e seu aluno (pelas palavras de Augusto), jamais esquecendo da amizade dedicada a nós.

Grato (a) por tanta plenitude, orientadora e amiga, Monique Nuijten.

Augusto Antonio Campelo Cabral / Mestre (2004) e Doutor (2013) em Geografia pela UFPE / Professor da rede estadual de Ensino de Pernambuco, Brazil / acampelocabral@gmail.com
Monica Cabral / amiga e foi auxiliar nas pesquisas de campo da professora Monique junto aos informantes-chave na comunidade Chão de Estrelas – Recife/PE, Brazil

36. MIRAR COMO MONIQUE NUIJTEN

Ernesto Martínez Fernández

Abril de 2016. Un estudiante de doctorado sube las escaleras del *Leeuwenborch*. Se dirige a uno de los despachos docentes para conocer a la que será la supervisora de su estancia de investigación en la Universidad de Wageningen. Está tenso. Su tesis se encuentra en un impás y, en cierta manera, espera encontrar respuestas en uno de los 'templos' de la antropología del desarrollo. Aunque intenta darse ánimos, le acecha la inseguridad. ¿Son sus preguntas pertinentes? ¿Ha sido su investigación de campo suficientemente productiva? ¿Qué pensarán de su trabajo personas cuyos apellidos está cansado de ver en prestigiosas publicaciones?

Llama a la puerta entreabierta del despacho. Con una sonrisa, la profesora lo invita a sentarse. De manera muy cercana, ella se interesa por su viaje hasta Wageningen, por su instalación e incluso le ofrece consejos para adquirir una bicicleta económica. En segundos, las tensiones empequeñecen. La calidez y humildad de ella marcan el resto de una conversación llena de sintonía en términos académicos y también políticos. El estudiante percibe rápidamente la excepcionalidad de la persona que tiene delante y sabe desde ese mismo instante que, más allá de la duración de su estancia en los Países Bajos, nunca querrá perder el vínculo con ella. El afortunado estudiante era yo y esa profesora llena de magnetismo Monique Nuijten.

Tres años después de esa estancia defendí mi tesis. No tengo ninguna duda de que dicha etnografía habría sido muy diferente sin esos meses y sin la participación de Monique. De su mano profundicé en los enfoques orientados hacia los actores y las prácticas. También revolví la bibliografía antropológica sobre el Estado y me perdí en Stuart Hall para luego encontrarme. Unos ámbitos que yo exploraba de manera autónoma, sin sugerencias u orientaciones concretas de Monique sobre determinadas publicaciones o temáticas de interés. En su lugar, eran sus reflexiones sobre mi trabajo y, sobre todo, el impacto que me produjo su forma de mirar a la realidad social lo que me impulsaba.

A mi juicio, esa óptica tiene mucho que ver con una antropología netamente microsocial, que se interesa por las personas reales y por las relaciones reales entre personas, sin pre-juicios ni excesos teóricos. Esa óptica tiene también que ver con un profundo respeto hacia los interlocutores con quienes estudiamos. Y, relacionado en parte con ello, con el compromiso, el de una antropología que no se reduce meramente a un campo (de fuerza) académico, sino que debe ser una herramienta de crítica para el cambio social. Toda su trayectoria investigadora da buena muestra de ello y véase como el mejor de los ejemplos su trabajo sobre '*grassroots politics*' en el seno de la Plataforma de Afectados por la Hipoteca en España.

Mal sabía yo cuando subía por primera vez aquellas escaleras que el encuentro que me aguardaba me ayudaría a ordenar tan claramente mis ideas en relación a la práctica y la practicidad de la antropología. Mirar desde el rigor, el cuidado y el compromiso de forma entrelazada – en una palabra, *mirar como Monique* – se convirtió, desde ese momento, en un anhelo que ya es parte de mí mismo.

Ernesto Martínez Fernández / Lecturer, Department of Social Anthropology, Psychology and Public Health, Pablo de Olavide University, Spain / emarfer2@upo.es

37. SHARING OFFICE

Regnar Kristensen

I cannot say my reaction was of the more cheerful ones when I was told that I had to share my office with a Dutch scholar. I was longing for peace. I was in the final period of writing-up my PhD dissertation at the Department of Anthropology at Copenhagen University. I even moved out of my house to save my wife from my shifting moods and 'non-presence'. And now I should share my office with somebody I did not know.

That was a decade ago.

What I remember clearly was Monique Nuijten's warm and friendly smile when I entered my office. A little shy I first thought but it did not take her more than a few minutes to change my initial irritation. She had shortly before been on fieldwork in Recife, Brazil, and had been granted a six month year visiting research stay at Copenhagen University. Very soon I felt being the luckiest man on earth. She asked me about my research on saints and security in Mexico City with genuine interest, laughed about my many anecdotes from my field site, and commented critically on whatever fluffy thoughts I had in my head. It was a miracle. What more could I have asked for in this narcissistic period of proving one's qualities as an anthropological researcher? Retrospectively I feel sorry for Monique to have had to cope with me in this period.

Slowly I learned more about her professional work. As so many others, I was impressed by her concept of development as a hope-generating machine; something that made much sense to one, who had lived in Mexico. 'Ah, that explains…', I remember thinking with astonishment.

As the weeks passed I missed her company if for some reason she was out of office.

We shared a fondness for long meticulous ethnographic research and conceptual thinking, and cheered all kind of writings where macro social realities didn't overshadow the narratives of the humans entrenched in those; be it in land reforms in Mexico, slum-upgrading in Brazil, or as in my case, clandestine economies in Mexico City. We shared loads of fieldwork anecdotes, some humoristic, others more tragic. I remember wondering what on earth made a woman like Monique conduct fieldwork under so difficult conditions, be it in the Spanish Pyrenees studying (male) shepherds far out on the mountain, farmers in remote *ejidos* in Mexico, or people living in poor urban quarters in a Northern Brazilian city. I think this shows a lot about Monique and the drive she has. Never afraid of entering difficult places to ask difficult questions. Her many publications shows her professional engagement with the people she does research with, always questioning the power structures behind the social inequalities and injustices she encounters.

Many cups of coffee and joyful discussions later things ended. I submitted my PhD and Monique returned to Holland. Before we exchanged farewells, I asked her to be part of my PhD committee. I knew at this point that I could not have asked for a better academic opponent. Theoretically sharp, innovative, knowledgeable on Mexican matters, and with this steel will behind her friendly and warm approach.

She accepted.

I might had feared my opponents' critique but none hit me harder than that of Monique. Respectfully she told me that she would have liked to know more about my interlocutors in my dissertation. What did they do for living, what were there anxieties, hopes and desires? In other words, where were the hundreds of anecdotes we had discussed and exchanged during the six month we shared office? Why was it not reflected in my PhD? Her critique pulled the carpet from under my feet. It moved me so much that I rethought my whole engagement as an anthropologist and the way I write ethnography. None of my later ethnographic works would have materialised without this subtle 'decapitation' of my dissertation.

This anecdote is what I believe characterises Monique most: besides being theoretically sharp and innovative, she is a genuine scholar, capable of listening and understanding her fellow anthropologists (and students, I suppose) in the same outstanding generous manner as she understands and cares for the people she works with.

Regnar Kristensen / Anthropologist affiliated with the Department of Cross-Cultural and Regional Studies, University of Copenhagen, Denmark / regnarkristensen@hotmail.com

38. THE SKILL OF WRITING OBSERVATIONS: WHAT MONIQUE TAUGHT ME

Edwin Rap

Monique Nuijten played an important role in my development as an ethnographic researcher. As my field supervisor, she taught me something that seems a minor matter, but it means the world to me. I am under the impression that many young researchers today do not have the time to capture, or fail to be taught, this basic ethnographic skill. In 1992, Monique and her family received me in their home in Autlán de Navarro, Jalisco in Western Mexico, near the *ejido* (land reform community) where she was doing her PhD research on land issues. I was a young irrigation student with broad and ever-expanding interests, who had ventured into many different subjects related to water management. In short, I was all over the place.

I arrived in Mexico with a research proposal trying to bring together the disciplinary perspectives of irrigation management and rural development sociology. For this purpose, I found Monique's use of the concept of 'organising practices' interesting, especially the focus on organising as a verb, which brings out the patterning processes in organisational practices and relates to power and politics in sometimes unpredictable ways. This implied a need for detailed ethnographic study of everyday organisational practices, also in irrigation management, rather than taking formal structures, managerial ideals, policy models, or legal documents as self-evident starting points for the analysis. My intention was to study those organising practices in Autlán-El Grullo's irrigation system in the wider context of neo-liberal irrigation policies. Monique arranged a visit to the office of the newly established Water Users' Association, where I got to know several people and which provided me legitimate access to the organisation.

To do fieldwork and to improve my Spanish, Monique inquired about where I could live with a family of *ejido* farmers, who also had irrigated land with sugarcane. She contacted a young family in the nearby village of El Chante, Carmen and Pedro with a young daughter and expecting a son. I was 'adopted' by them and went to live in El Chante. Carmen cooked for her family including myself three times a day and over these meals I learned my first Mexican Spanish. Part of learning the language was discovering a characteristic Mexican humour. I remember sitting at a large family table in front of Pedro, while Carmen was making the tortillas by hand on the stove in the kitchen. I was studying how Pedro was using the tortillas to scoop up the beans from his plate, while trying to have a conversation with him. On the table always sat a big bowl of chilis, present at many Mexican dinner tables. On one of the first days, Pedro pushed it towards my plate with a meaningful glance in his eyes. Out of politeness, I put a couple of chilis on my plate and Pedro showed me how to take a bite of it, by acting as if he would do this himself. There was no way out it seemed to me, so I took a good bite of the *jalapeño* pepper. My mouth was slowly set on fire and I started sweating everywhere, which I tried to hide by wiping my face and drinking water. Meanwhile, Pedro was studying me closely. It must have been so funny to see this tall, blond, European novice in this predicament. Then he asked me: 'And?' I praised Carmen's wonderful food, but stubbornly refrained from mentioning the effect of this pepper on me. 'But the chilis, aren't they hot?'. 'Well, a little bit', I had to admit. 'Yes', he said, 'that's why I never take them…' Through partaking in this funny kind of initiation, I learned my first lesson as an outsider about entering a specific culture of power, in which masculinity figured prominently. The meaning(s) of this interaction took me years to figure out, let alone find the words to theorise. But a seed was planted.

These were some of the observations that I enthusiastically shared with Monique, the first time I sat down with her in her office. She was pleased and interested, I could see, but then she asked me the key question: 'Can I see your field notes…?' I immediately felt incredibly embarrassed, because I knew I could only show a few uncoordinated jottings in a folded, wet, and greasy notebook. I do not remember Monique's exact response, but the message was clear: 'You need to write this stuff down!' She showed me the importance of writing down observations of everyday life in the form of extensive field notes when they are still fresh in your mind. Monique made clear that elaborating seemingly irrelevant and unexpected, but striking and insightful, details requires a lot of time and commitment, but that this would pay off in the long run. This was what I needed. By learning this lesson, I found a focus that would make me an ethnographer. To this day, I carry my notebook around everywhere.

Edwin Rap / Independent researcher & adviser, Rap[id], the Netherlands / edwin.rap@gmail.com

39. MONIQUE NUIJTEN AS A HOPE GENERATOR IN CONTRARIOUS FORCE FIELDS

Oscar Salemink

Ever since the publication of her *Power, Community and the State: The Political Anthropology of Organisation in Mexico* (2003), Monique Nuijten is widely considered as one of the foremost and most original thinkers in what has been termed 'critical development studies'. In this painstakingly researched book she describes and analyses in meticulous ethnographic detail the encounter of *ejidatarios*, landless peasants, private landowners, state officials, surveyors, priests and other actors at the interface (cf. Long, 1989) of land and development politics. Two years later, she published a theoretical treatise on the 'force field approach' towards power which seeks to remedy the theoretical poverty of development studies by seeing it as embedded in relational power, with reference to various, intertwined institutions implicated in development (Nuijten, 2005).

Conceiving of development as a field of political practice and intervention was perhaps not new in the early 2000s, but in contrast with many earlier approaches, like Arturo Escobar's (1995) or Gilbert Rist's (1997), Nuijten eschewed the distant, abstract deconstruction of grand narratives and critique of big institutions for an approach that is grounded in local realities and life-worlds of people of flesh and blood. Seeing and empathising with the desire for development among her interlocutors on the ground, she could understand the attraction of narratives and discourses of development as a 'hope-generating machine' (Nuijten, 2003) – an enduring attraction that would survive frequent disenchantments. One could argue that it is this perennial attraction of the discourse of development for poor and deprived categories of people as catering to their desire for a better life, which makes their integration into the state market nexus and their subjection to capital possible, most egregiously brought out in the process that David Harvey (2005) called accumulation through dispossession. In that sense, Monique's work preceded by one decade more recent studies by Tania Murray Li (2014), Holly High (2014) and myself (Salemink, 2015a) about development as a 'field of desire for modernity'.

Monique Nuijten and I followed somewhat similar trajectories. Both of us have lived and worked for a longer period of time in the countries where we did our doctoral research – in Monique's case Mexico, in my case Vietnam – where we were both in our own ways and in different contexts intimately involved in development processes. After that period we both came back to the Netherlands and (re)joined academia there, using our respective experiences for critical but engaged reflections on development, power, economic (ownership) transactions and cultural expressions that place the so-called 'target groups' of development at the forefront. In a tiny country like Holland it is little wonder that Monique and I ran into each other, overlapping as editors of *Focaal – Journal of Global and Historical Anthropology* and as members (together with Philip Quarles van Ufford, Rüdiger Korff, David Lewis and David Mosse, to name a few) of the informal and irregular discussion group EIDOS (European Inter-University Development Opportunities Study-group) that sought to afford a free, safe and creative space for non-conformist thinking about development.

That period of substantive and theoretical convergence between Monique and myself in the 2000s culminated in the EIDOS conference 'The Ends of Development: Market, morality, religion, political theology?' that we organised together in 2008. The title of the conference was a pun on the critique of and disenchantment with development by understanding 'the end' in temporal terms, much like in Andrew Brooks' recent *The End of Development* (2017), which is advertised as a 'scathing indictment of the current development agenda, and an impassioned call for a new and radical approach to alleviating global poverty'. But we also played with the moralising meaning of 'ends' as goals, by 'question[ing] once again the stated aims of development – not in terms of targets, but in terms of its societal ideals. In other words, what are the visions of the *teloi* of development – as a project of a vision of secular modernity – in a re-enchanting world?' [emphasis in original].[12] That arguably more transcendental take on 'ends' was subsequently taken up in volumes like *Religion and the Politics of Development* (Bush *et al.*, 2015)[13] and *Political Theologies and Development in Asia* (Bolotta *et al.*, 2020).

In other words, the 2008 EIDOS conference combined Monique Nuijten's theoretical interests in conceiving of development as a political arena of desire, intervention and contestation, and my interests in conceiving of development in political theology terms as a form of 'quasi-religious conversion' (cf. Salemink, 2004). Although the conference generated ideas that seemed promising then (which is perhaps brought out by the fact that some of these ideas were taken up in some later publications by some scholars mentioned above), it felt like a failure to us. The reason for that was that at the time Monique's and my life trajectories converged again, but along more personal lines: we were both experiencing a downward spiral in our private lives for uncannily similar reasons, which impacted on our professional lives through distraction, distress, confusion and lack of energy. In other words, we were both for personal reasons unable to turn the ideas and debates brought up at 'The Ends of Development' conference into tangible publications – at least temporarily.

[12] Cf. Call for Papers 'The Ends of Development: Market, Morality, Religion, Political Theology?', An international EIDOS (European Inter-University Development Opportunities Study-group) conference. VU University Amsterdam, 19-20 June 2008. Issued in November 2007.

[13] I contributed a chapter titled 'The purification, sacralisation and instrumentalisation of development' to that volume (Salemink, 2015b).

Yet, meeting with, getting to know, working closely with and going down simultaneously with Monique was in hindsight one of the best experiences in my life. When I was down-and-out and incapable of thinking straight in the late 2000s, Monique held up a mirror, held out a helping hand, and offered support rooted in the profound wisdom that comes with a scarred life. In return, she often said the same of my role in her life, so the image of that period that I have in my mind is that while both of us were lacking solid footing, we pulled each other by our hair out of the morass that both our lives constituted at the time. Both of us fortunately recovered and went our own ways professionally and personally. Substantively and theoretically, I moved away from development as a theme, towards an anthropology of the 'secular sacred' as materialised in cultural in practices around cultural heritage, art and museums and their political-economic interconnections, whereas Monique developed her interests in organisations, law, property, slums and corruption as contested political processes. Personally, I 'resolved' my situation in 2011 by moving to Copenhagen, coincidentally at a time that Monique began a relationship with a Danish man based in Copenhagen.[14] Spatially separated, Monique and I continued to enjoy a close friendship, but more recently our life trajectories took on an uncannily similar curve again when only few months apart both of us were diagnosed with serious forms of cancer – which, of course, is the occasion for writing this text.

My personal experience with Monique gave me – apart from joy and mutual benefit – an invaluable insight, namely that concepts may to some extent be seen as extensions of their originators. Although calling someone a 'machine' is not especially respectful, Monique is in my eyes very much like a 'hope-generating machine', or at least a 'hope-generator'. She is someone who generously and selflessly spreads her ideas, wisdom and generative warmth around, and evokes hope in others – in more ordinary mortals like myself. Even in the face of a mortal illness she remains positive, cheerful and humorous. At the same time, while she is keenly aware of life as conditioned by 'force fields', she comes across as a powerhouse herself, although never in an overbearing manner, or in the form of intellectual machismo (M/F). Assuming that we operate in 'force fields', Monique is a strong magnet deriving her strength and power from attracting people in a positive sense.

If indeed some of Monique Nuijten's main, innovative concepts – development as 'hope-generating machine', and 'force field', amongst others – can be interpreted as extensions of her personality, perhaps we may dare to cautiously and hypothetically test this intuitive insight more generally. Perhaps Karl Marx's focus on labour, capital and alienation betrays his reportedly irritable, suspicious and often sarcastic character (something that biographers attribute to his marginal position as a Jew and his ill health). Perhaps Antonio Gramsci's interest in the working of hegemony was rooted in his failure to achieve that himself, as he was able to contemplate while in prison.

[14] I refer to Finn Stepputat, of course.

Perhaps Michel Foucault's reduction of social life to discursive formations is in line with a cerebral element in his character (although he did enjoy S/M). Perhaps Gilles Deleuze's philosophical preoccupation with desire was inspired by his putatively dull academic's life. Perhaps the emphasis on flat, a-historical networks in Bruno Latour's Actor-Network-Theory mirrors Latour's own superb networking skills and flatness, and-so-on and-so-forth. Young scholars these days are told that they should develop their own, new concepts in order to make it in academia, with the ultimate hallmark of success being that one's last name is turned into an adjective: Marxist, Gramscian, Foucaultian, Deleuzian, Latourian, Bourdieuan, etc. The ability to turn a last name into an adjective has, of course, to do with the phonetic quality of the name ('Marxist' or 'Marxian' simply sounds better than 'Bourdieuan'), but more importantly with the gender of the thinker: few female thinkers had their last names turned into adjectives.[15] What seems to work well for white, male, continental European philosophers and also contemporary French thinkers, is difficult for most others.[16]

Monique has the disadvantage of having a name that is phonetically difficult to render as an adjective – just try to pronounce 'Nuijtenian' – and in an academic world marked by intellectual machismo she has the double disadvantage of being a woman. But perhaps we should overcome our hesitation and use the adjective Nuijtenian freely, in order to not just honour Monique's scholarly accomplishments as a 'concept-generating machine', but as a scholar whose concepts are grounded in her grounded ethnographic experience and interwoven with her personality. Seen from that latter angle, I propose that 'Nuijtenian' comes to mean the scholarly inspiration in combination with the generosity and warmth that characterise Monique Nuijten as a person.

Oscar Salemink / Professor of Anthropology, University of Copenhagen, Denmark / o.salemink@anthro.ku.dk

References

Bolotta, G., Fountain, P. and Feener, M. (eds), 2020. Political theologies and development in Asia: transcendence, sacrifice, and aspiration. Manchester: Manchester University Press.

Brooks, A., 2017. The end of development: a global history of poverty and prosperity. London: Zed Books.

[15] There are some exceptions, like Hannah Arendt ('Arendtian'), Simone de Beauvoir ('Beauvoirian'), Julia Kristeva ('Kristevian'), Chantal Mouffe ('Mouffian'), Judith Butler ('Butlerian'), but some names do not lend themselves to such inflections, like 'Nussbaumian' (after Martha Nussbaum). Even though not impossible, the conversion of the last names of great women thinkers into adjectives is highly uncommon, suggesting that this is a gendered phenomenon.

[16] As for me, I do not think that I ever coined a new concept, which perhaps makes it more difficult to assess my personality through my concepts.

Bush, R., Fountain, P. and Feener, M. (eds), 2015. Religion and the politics of development. Basingstoke: Palgrave MacMillan, pp. 35-60.

Escobar, A., 1995. Encountering development: the making and unmaking of the third world. Princeton: Princeton University Press.

Harvey, D., 2005. A brief history of neoliberalism. Oxford: University Press, Oxford.

High, H., 2014. Fields of desire: poverty and policy in Laos. Singapore: NUS Press.

Li, T.M., 2014. Land's end: capitalist relations on an indigenous frontier. Durham, NC: Duke University Press.

Long, N. (ed.), 1989. Encounters at the interface: a perspective on social discontinuities in rural development. Wageningen: Wageningen Agricultural University.

Nuijten, M., 2003. Power, community and the state: the political anthropology of organisation in Mexico. London: Pluto.

Nuijten, M., 2005. Power in practice: a force field approach to natural resource management. Journal of Transdisciplinary Environmental Studies 4(2): 3-14.

Rist, G., 1997. The history of development: from western origins to global faith. London: Zed Books.

Salemink, O., 2004. Development cooperation as quasi-religious conversion. In: Salemink, O., Van Harskamp, A. and Kumar Giri, A. (eds) The development of religion, the religion of development. Delft: Eburon, pp. 121-130.

Salemink, O., 2015a. Revolutionary and Christian ecumenes and desire for modernity in the Vietnamese highlands. The Asia Pacific Journal of Anthropology 16(4): 388-409.

Salemink, O., 2015b. The purification, sacralisation and instrumentalisation of development. In: Bush, Fountain, P. and Feener, M. (eds) Religion and the politics of development. Basingstoke: Palgrave MacMillan, pp. 35-60.

40. THE PLEASURE OF NARRATING LIVES: MONIQUE AS A PASSIONATE ETHNOGRAPHER

Pieter de Vries

Those who know her will ascertain that Monique's passion for anthropology is driven by a fascination in finding precise answers to a set of recurrent questions. In this essay I provide examples of her passion for research. I concentrate on our shared (life) experiences and her early field work in the Pyrenees and Mexico.

Monique and I became close friends in 1985 when we both were student assistants at the Department of Non-Western Agrarian Sociology, chaired by Professor Norman Long. Unimaginable now, she was very shy in public. In class and in seminars, she rarely spoke. Yet, she was an excellent presenter; passionate and sharp. That was striking for many, like me, who knew her to be a very sociable person. I remember that she managed to obtain permission to borrow as many as 20 books from the library at a time. Her student room was piled high with books, most of which were ethnographies, devoured for possible cases and explanations.

Monique's reputation as a promising scholar began with her early research in the Spanish Pyrenees on transhumance shepherds which she did for her BSc degree. Together with Willemien Brooijmans, she set out to document the summer travels of shepherds to higher grounds (the two would accompany shepherds for days in pursuit of this story). She recounted the solitude of these men, some of whom would never find a life partner in villages that were being abandoned, as young women relocated to the city in search of work. Monique gathered extensive data on the multi-layered agricultural strategies and the complexity of property relations in the region. Strikingly, some families were becoming quite wealthy buying land as well as houses in the village, which begged the question: what for? If there seemed to be no future in the village, then why were some families buying up property? Men accumulated capital with a view to getting a spouse, but found it difficult to start a family. Migrant women lived afar, but maintained close relationships with the village. Households were fragmented and yet there was a strong sense of commonality.

One of the questions Monique raised was whether this phenomenon was an inexorable effect of modernisation. No, was her answer, drawing on Long's concept of differential responses to change. Incorporation theory was popular at the time, but she was critical of it and preferred to look at the relationship between the local and the global from a historical perspective. What fascinated her was the ability of people to partake in different worlds (e.g. the village in the Pyrenees, the city) and fields of activity (e.g. pastoralism, agriculture, commerce). The dynamics of village life, migration, shifts in property relations, and the historical relationship between the local and wider power fields was what sparked her interest.

Norman Long was impressed by her work, in particular the detail with which she documented and analysed different fields of activity. Her approach to research fit snugly with the Manchester School tradition: looking at processes as multifarious and non-linear, a strong interest in unexpected responses, the contingency of everyday life, and the making of local worlds. Giddens' structuration theory was popular at the time, but Monique was doubtful about its capacity to bridge the agency/structure divide. She actively participated in the development of the actor-oriented approach, but she was not a follower, as she had her doubts about the agentive capacities of actors to shape their lives in all circumstances and moments. At the same time, she had a keen interest in organisation theory, legal anthropology and the ethnographic case study method.

In 1987 Monique started doing her MSc field research in Mexico, following her previous interests in the Pyrenees. This was a part of Norman's project on irrigation and planned intervention in Western Jalisco, Mexico, in two valleys where sugar cane and agro-export production for the USA (including marihuana) were very strong. Western Jalisco is the typical cowboy (*ranchero*) area as depicted in Mexican films, characterised by deep Catholicism, conservatism, individualism, ruggedness and a good deal of racism against indigenous people. Monique went to live in the village of Lagunillas in the house of one of the founding families of the *ejido* (a communal form of smallholder agriculture). This was a typical transnational family, involved in maize production for local markets, cane production for the national sugar market, and illegal migration to the USA (where half of the village lived).

In doing research, Monique took on the role of a daughter in the family, following the same rules as other unmarried daughters. There was a lot of bantering, gossip, and partying (at the bullfights), especially during the dry season when labour requirements are low. But Monique also visited the fields (which was uncommon for a young woman), and she wrote down everything she observed about the extended family. She started looking at the family as a field of activity shaped by a conservative ideology that did not correspond to the practice of everyday decision-making (we referred to this as 'cognitive dissonance' at the time, before poststructuralist discourse theory). Notably, Monique's 'village mother' came from a traditional family and was very well-connected. She had a strong character and was instrumental in the family decision to migrate to Los Angeles; an unexpected and sudden decision that fascinated Monique.

Monique's MSc research was highly ambitious. Again, she was interested in differential forms of integration in wider fields of activity, and thus, the history of migration, differential responses to change, and the contingency of decision-making. She traced the history of these activity fields which extended into the USA. One question that obsessed her was why so many men chose for temporary migration, and then returned to the village to grow maize; a crop that was not profitable. What was it that motivated them to engage, yearly, in these forms of illegal migration? Women, on the contrary, adapted more easily to the way of life in the USA. There was much demand for their work in the service sector and they enjoyed the conveniences of city life.

Strikingly, Monique did not find feminist theory very convincing for her research, especially the critique of patriarchy as a dominant structure of domination. Patriarchy, in her view, was highly flawed and inconsistent, even ideologically. Younger generations of men were unable to be reliable providers like their predecessors. The result was an increase in masculine violence, including domestic abuse. Worse than the symbolic violence of the family patriarch, was the pathetic, but very real violence of men frustrated by the empowerment of women due to migration. The resentment triggered by these changes spurred more violence against women when drug gangs came to dominate the area.

Yet, Monique was very interested in gender dynamics at the village level: contrasting the performance of machismo to the increased role of women in decision-making processes. She had a couple of male key informants, whose intelligence she appreciated. One of her favourite informants was Javier[17], an *ejidatario* (member of the ejido) who never migrated to the USA, and who cherished maize agriculture. He did not belong to one of the founding, powerful families and avoided local *ejido* politics and disputes. His two daughters had moved to the city of Guadalajara and he did not have heirs interested in his land. So, he knew that he belonged to the last generation of traditional farmers in the village and he accepted that as a matter of fact. The future, he said, belonged to the youth and especially women.

[17] Not his real name; he appears on the front cover with Monique and his wife.

So, Monique sympathised with the fate of men, their solitude, their impotence, in a violent and *machista* world, while remaining highly interested in the silent power of women who were gaining increasing control over their sons and daughters through migration. At the same time, she was weary of overly romantic analyses. She was a realist, always attentive to the local worlds and frames of interpretation of the people she studied, and to the role of larger structural forces that divided families and communities and that imposed new forms of violence on the poor, especially women and the younger generations.

Back in the Netherlands in 1990, Monique started working on a PhD research proposal, taking as her topic the *ejido* as a land-holding institution. At the time there was much talk about the privatisation and liberalisation policies of the *ejido*. As Mexico prepared to enter the North American Free Trade Agreement in 1991, President Carlos Salinas de Gortari declared the end of *ejid*o land distribution and allowed existing *ejidos* to rent out and sell land, thus ending land reform in Mexico. This was a highly ideological topic, and many predicted that this was the death knell for smallholder farming in Mexico. Surprisingly, people in Lagunillas were far less impressed by the possible impacts of these policies than academics. They welcomed these changes as a recognition of how things actually worked. Monique's research proposal (later published in Norman and Ann Long's edited volume *Battlefields of Knowledge*, 1992), set out to study these policies by inquiring into the history and dynamics of property relations.

She started her proposal with an anecdote concerning her search for a formal document: a list of *ejidatarios* with an official map of their belongings. After pursuing the *ejido* commissioner for months, he invited Monique for a soft drink and started to describe every ejido properties, detailing the borders. He did this by rote. The point was that there was no map. That was the moment she learnt that *ejidatario* smallholders did not read official maps, yet they produced detailed mental relational maps that combined local knowledge about family inheritance with the physical boundaries of plots.

Back in Lagunillas, Monique started elaborating genealogies of land. She did that through a combination of archival research and oral histories. Central in her research were issues of sale, renting and inheritance, in relation with changes in the *ejido* law and shifting family dynamics in the village. She showed that property was encumbered by different sets of claims and expectations. Inheriting land implied retaining a close relationship with the village, something that was difficult to sustain by migrants. Would the liberalisation of the *ejido* law lead to further land concentration by *caciques* (strongmen) supported by private capital? Her research showed that there had always been a market of land in the village, but that market forces had always been mitigated by the existence of normative frameworks – *de facto* a kind of non-state law – that regulated land transactions. Further, the dominance of local strongmen was in true decline, given the increasing importance of transnational livelihoods.

While doing research on the workings of the agrarian bureaucracy Monique realised the opaque and speculative character of the force-fields in which *ejidatarios* operated. She explored the case of a large plot that had never been distributed (and was still in the hands of private landholders), which included accompanying a commission of *ejidatarios* as they sought justice through numerous state offices. This quest inspired her to delve more deeply into the workings of bureaucracy as a machine devoid of any kind of formal rationality

Monique also became interested in corruption; not in formal terms as the abuse of bureaucratic power, but in its performance, its pleasurable and speculative side. She argued that the discourse of corruption enables state subjects to engage in spectacular imaginations of the state. In doing so she engaged with the bureaucrats' discourse, imagination and performance of corruption. For engineers in the agrarian bureaucracy it was acceptable to be a little corrupt, but they criticised colleagues who were 'very corrupt', thus constructing moral criteria for dealing with *ejidatarios*. Migrants from Lagunillas complained that there was no freedom in the USA, as you cannot negotiate situations with police, thus constructing an image of the state there as being overly repressive and intolerant. Corruption in her work is thus performative, it provides possibilities for imagination and is enjoyable.

She applied this perspective to the analysis of conflicts and disputes that were never resolved, and hence brought forward an idea of the state as an opaque and aloof apparatus bolstered by the fiction that there is a central locus of power where decisions are made. This led her to analyse the workings of the state bureaucracy as a hope-generating machine; an apparatus that promises rules and closure, but in fact, operates and rules through procrastination and inefficiency, converting agentive citizens into patients of the state.

Norman, her supervisor, was initially doubtful about her approach which he saw as a flirtation with semiotics (the science of signs), and thus implicitly a structuralist deviation. Yet, he did propose the thesis for a *cum laude*. Two renowned Latin Americanists were on the committee: Gavin Smith and John Gledhill. I remember that John, who is stingy with praise, was deeply impressed by the quality of the ethnography. Interestingly, the defence revolved very much around ethnographic issues with Gavin asking detailed questions about several characters in the ethnography before formulating a theoretical question.

In the *judicium* (assessment) Norman announced that she was indeed granted a *cum laude*. This was completely out of the blue for Monique – no one had informed her beforehand that it was even a possibility. Yet, it was hardly surprising as she had achieved a *cum laude* for all of her graduations.

In 1994 I got a job at the Colegio de Michoacán and we moved with our twin daughters to Zamora. Shortly after, the department of anthropology appointed her as a research fellow. I remember José (Pepe) Lameira, one of the luminary anthropologists of

COLMICH telling me how impressed he was by this young, charming woman and her analytical capacities and passion when talking about her research, and the *ejido* in particular. Pepe commended the courage of young foreign researchers who were not afraid to live in distant and insecure rural areas and (to our amusement) commented that she did it all with a delightful French accent! Curiously, Monique became known there as a sort of post-modernist, as she did not believe in the grand narratives of progress and revolution. She found it a funny characterisation as she was absolutely not interested in this kind of debate. What interested her was the fate of ordinary people; their intimate, little worlds, their hopes and disappointments. She had little faith in the prospect of grand ideas or designs for changing the world.

We see a consistent pattern in Monique's ethnographic perspective. It is about people and their construction of intimate, yet unbounded, worlds. She is quite pessimistic about notions such as progress or development, but also about the capacity of people to shape their own lives in a strategic, intentional way. She has been interested in the imaginary, spectacular side of power, its labyrinthine and opaque features. Social process for her was always incomplete and unfinished. As she liked to say to students: 'never look for closure when writing an ethnographic case study. You never know how the story may shift in surprisingly unexpected ways later.'

What, then, makes Monique a great anthropologist in a neoliberal academic environment where ethnography is increasingly under threat. Theories surge and decline, and although always conceptually solid, Monique was never married to a particular theory, school, or thinker. Although interested in gender dynamics she never abided with feminist theory. Although critical of development interventions and the hubris of modernist aspirations to change people's worlds she never adhered to a post-development perspective. Although always on the side of ordinary people and attentive to their capacity to shape their worlds, she avoided the romanticism of resistance, and the entrepreneurialism of actor-oriented approaches. She was acutely aware of the risks, injustice and humiliations inflicted on the powerless. Although she was highly impressed by poststructuralist theory, she never adhered to fashionable ideas of disciplinary power or biopolitics. Although interested in the performative and imaginary side of power, she never indulged in Lacanian theories on enjoyment and desire. Neither was she interested in fashionable theories of transgression.

Monique is conceptually eclectic, always drawing on contemporary ethnographic debates. She draws inspiration from literature, and 'personal interest stories'. Reading her work gives the reader a sense of how people manage their lives, including the way in which imagination and story-telling shape organizing practices. Emanating from her writing are the strong relationships established with people among whom she did fieldwork, the pleasure she derived from telling their stories, and the ways she made their happiness and suffering her own. It is this strong commitment to the fate of common people and her capacity to theorise it that makes her work important and long lasting. Not only because she did contribute to the shaping of a disciplinary field, but also because future generations – especially in the South – will be able to identify with the characters in her ethnographies.

I end this essay with immense gratitude and appreciation to a wonderful researcher and companion.

Pieter de Vries / Assistant Professor, Sociology of Development and Change Group, Wageningen University, the Netherlands / pieter.devries@wur.nl

References

Long, N. and Long, A. (eds), 1992. Battlefields of knowledge: the interlocking of theory and practice in social research and development. Routledge.

ACADEMIC PUBLICATIONS MONIQUE NUIJTEN

Books, edited volumes and theme issues of journals

2019 (edited with Maritza Bode Bakker and Jeroen Bok). *Lucha y Esperanza: Testimonios de participantes en la PAH* [Struggle and hope: Testimonies from participants in the Platform of Mortgage Victims]. Wageningen: Wageningen University & Research.

2013 (edited with Sian Lazar). Special Issue: Citizenship, the Self, and Political Agency. *Critique of Anthropology* 33(1).

2008 (edited with Gerhard Anders). *Corruption and the secret of Law: a legal anthropological perspective.* Farnham UK: Ashgate.

2003 *Power, community and the state: the political anthropology of organisation in Mexico.* London: Pluto Press.

Articles

2018 (with Maritza Bode Bakker). 'When breaking you make your soul dance' Utopian aspirations and subjective transformation in breakdance. *Identities* 25(2): 210-227.

2018 (with Martijn Koster, Pieter de Vries and Augusto Antonio Campelo Cabral). Regimes de ordenação espacial no Brasil: A fusão de neoliberalismo, populismo de esquerda e visões modernistas na urbanização de favelas no Recife [Regimes of spatial ordering in Brazil: The fusión of neoliberalism, leftist populism and modernist urbanist visions of favelas in Recife]. *Caderno CRH* 31(82): 59-73.

2016 (with Martijn Koster). Coproducing urban space: Rethinking the formal/informal dichotomy. *Singapore Journal of Tropical Geography* 37(3): 282-294.

2016 (with Lynn van Leerzem and Pieter de Vries). Being *Negro* in Recife Brazil: A political aesthetics approach. *Ethnos: Journal of Anthropology* 81(5): 865-887.

2015 Political activism and the PAH (Platform of Mortgage Victims) in Spain: The significance of Franz von Benda-Beckmann's work for the study of power, political agency and legal pluralism at the grassroots. *Journal of Legal Pluralism and Unofficial Law* 47(3): 476-492.

2013 The Perversity of the 'Citizenship Game': Slum-upgrading in the urban periphery of Recife, Brazil. *Critique of Anthropology* 33(1): 8-25.

2013 (with Sian Lazar). Introduction: Citizenship, the self and political agency. *Critique of Anthropology* 33(1): 3-7.

2012 (with Martijn Koster). From preamble to post-project frustrations: The shaping of a slum upgrading project in Recife, Brazil. *Antipode* 44(1): 175-196.

2012 (with Martijn Koster and Pieter de Vries). Regimes of spatial ordering in Brazil: Neoliberalism, leftist populism and modernist aesthetics in slum upgrading in Recife. *Singapore Journal of Tropical Geography* 33(2): 157-170.

2012 (with David Lorenzo). 'Né pour être esclave': Migration transnationale des bergers d'une comunidad andine [Born to be a slave: Transnational migration of shepherds in an Andean community]. *Anthropologica* 51: 67-80.

2009 (with David Lorenzo). Ruling by record: The meaning of rights, rules and registration in an Andean Comunidad. *Development and Change* 40(1): 81-103.

2007 Recht in 'Ontwikkeling': De impositie van de 'Rule of Law' agenda in ontwikkelingslanden [Law in 'development': The imposition of the 'rule of law' agenda in developing countries]. *Recht der Werkelijkheid* 2007(2): 51-56.

2005 Power in practice: A force field approach to power in natural resource management. *Journal of Transdisciplinary Environmental Studies* 4(2): 3-14.

2004 Peasant 'participation', rural poverty and the State in Western Mexico. *The Journal of Peasant Studies* 31(2): 181-209.

2004 Between fear and fantasy: Governmentality and the working of power in Mexico. *Critique of Anthropology* 24(2): 209-230.

2003 Re-framing the field of action: The anthropologist between broker and activist. *Anthropology in Action* 10(2): 4-12.

2003 Land and the law: Towards a force field approach in legal anthropology. *Journal of Legal Pluralism and Unofficial Law* 48: 163-183.

2003 Family property and the limits of intervention; the Article 27 reforms and the PROCEDE program in Mexico. *Development and Change* 34(3): 475-497.

2003 Illegal practices and the re-enchantment of governmental techniques. *The Journal of Legal Pluralism and Unofficial Law* 35(48): 163-183.

2002 (with Kirsten Appendini). El papel de las instituciones en contextos locales [The role of institutions in local contexts]. *Revista de la CEPAL* (abril 2002): 71-88.

2002 Organizing the peasants: Participation, organization and the politics of development in a Mexican government program. *Anthropologica* 44(2): 223-235.

2000 (with Gemma van der Haar). The Zapatistas of Chiapas: challenges and contradictions. *European Journal of Latin American and Caribbean Studies* (68): 83-90.

Book chapters

2018 (with Finn Stepputat). Anthropology and the enigma of the state. In: *Handbook of political anthropology.* Harald Wydra and Bjørn Thomassen (eds.). Cheltenham UK: Edward Elgar Publishing, pp. 127-144.

2012 Beyond modernist thinking: Unmasking the myth of collective organization in the development debate. In: *Against the grain: advances in postcolonial organization studies.* Anshuman Prasad (ed.). Copenhagen: Copenhagen Business School Press, pp. 177-199.

2010 (with David Lorenzo and Pieter de Vries). Terre, discipline en citoyennété différenciée dans une communauté des hautes terres du Pérou [Land, discipline and differentiated citizenship in a community in the highlands of Peru]. In: *Politique de la terre et de l'appartenance: Droits fonciers et citoyenneté locale dans les sociétés du Sud* [Politics of land and belonging: Land rights and local citizenship in Southern societies]. Jean-Pierre Jacob and Pierre-Yves le Meur (eds.). Paris: Karthala, pp. 403-428.

2010 Legal responses to cultural diversity: Multi-ethnicity, the State, and the law in Latin America. In: *Cultural diversity and the law: State responses from around the world.* Marie-Claire Foblets, Jean-François Gaudreault-DesBiens, Alison Dundes Renteln (eds.). Brussels: Brussels Bruylant, pp. 235-254.

2009 (with David Lorenzo). Ritual and rule in the periphery: State violence and local governance in a Peruvian comunidad. In: *Rules of law and laws of ruling: on the governance of law.* Franz von Benda Beckmann, Keebet von Benda Beckmann, and Julia Eckert (eds.). Ashgate: Aldershot and Burlington, pp. 101-124.

2009 (with David Lorenzo). Peasant community and territorial strategies in the Andean Highlands of Peru. In: *Spatializing law: An anthropological geography of law in society.* Franz von Benda Beckmann, Keebet von Benda Beckmann and Anne Griffiths (eds.). London and New York: Routledge, pp. 31-58.

2009 (with David Lorenzo). Duenos de todo y de nada! (owners of all and nothing): Indian territories in the Central Andes of Peru. In: *The rights and wrongs of land restitution: Restoring What Was Ours.* Derick Fay and Deborah James (eds.). London and New York: Routledge, pp. 185-208.

2007 (With Gerhard Anders). Corruption and the secret of law: An introduction. In: *Corruption and the secret of Law: A legal anthropological perspective.* Monique Nuijten and Gerhard Anders (eds.). Farnham, UK: Ashgate, pp. 1-24.

2006 (with David Lorenzo). Moving borders and invisible boundaries: a force field approach to property relations in the commons of a Mexican *ejido.* In: *Changing properties of property.* Franz von Benda Beckmann, Keebet von Benda Beckman and Melanie Wiber (eds.). Oxford and New York: Berghahn Books, pp. 218-242.

2005 (with Jilles van Gastel). The history of the 'good governance' and 'ownership' agenda at the Dutch Ministry of Development Cooperation. In: *The aid effect: Giving and governing in international development.* David Mosse and David Lewis (eds.). London : Pluto Press, pp. 85-105.

2005 Transnational migration and the re-framing of mormative values. In: *Mobile people, mobile law: Expanding legal relations in a contracting world.* Franz von Benda Beckmann, Keebet von Benda Beckmann, Anne Griffiths (eds.). Aldershot, UK: Ashgate, pp. 51-68.

2004 The 'hope-generating machine' and the positionality of the anthropologist. In: *The development of religion, the religion of development.* Oscar Salemink, Anton van Harskamp and Ananta Kumar Giri (eds.). Delft : Eburon Academic Publishers, pp. 51-58.

2004 Governance in action: Some theoretical and practical reflections on a key concept. In: *Globalization and development: Themes and concepts in current research.* Don Kalb, Wil Pansters and Hans Siebers (eds.). Dordrecht, Boston, London: Kluwer Academic Publishers, pp. 103-130.

2003 (with Pieter de Vries). Cultural difference and the evocation of otherness: Reflections on the (mis)use of the culture concept in Andean studies. In: *Imaging the Andes: Shifting margins of a marginal world.* Ton Salman and Annelies Zoomers (eds.). Amsterdam: Aksant (CEDLA Latin America Studies 91), pp. 64-80.

2002 (with Pieter de Vries). Some reflections on the (mis)-use of the concept of culture in Andean studies. In: *Imaging the Andes: Shifting margins of a marginal world.* Ton Salman and Annelies Zoomers (eds.). Amsterdam: Het Spinhuis.

2001 Organisation and development: A practice perspective. In: *Resonances and Dissonances in Development. Actors, Networks and Cultural Repertoires.* Paul Hebinck and Gerard Verschoor (eds). Assen: Van Gorcum, pp. 139-152.

2001 Family property and the limits of Intervention: The effects of the Article 27 Reforms and PROCEDE in Mexico. In: *Los Mexicanos y la Tierra.* Center for Mexican studies. 94-109.

2001 What is in the land? The multiple meanings of land in the transnationalized Mexican village. In: *Land and Sustainable Livelihood in Latin America.* Annelies Zoomers (ed.). Amsterdam: KIT Publishers, pp. 71-92.

2001 The burden of kinship: Transnational networks and the inheritance of land in a Mexican family. In: *Kinship structures and enterprising actors: Anthropological essays on development.* Jens A. Andersson and M. Breusers (eds.). Wageningen: Ponsen & Looijen, pp. 55-76.

1998 Recuerdos de la tierra: luchas locales e historias fragmentadas [Memories of the land: Local struggles and fragmented histories]. In: *Las Disputas por el Mexico Rural. Transformaciones de Practicas. Identidades y Proyectos. Vol. II Historias y Narrativas.* Sergio Zendejas and Pieter de Vries (eds.). Zamora de Hidalgo: El Colegio de Michoacán, pp. 165-210.

1997 Memories of the land: Local struggles and fragmented histories. In: *México en movimiento; concierto mexicano 1910-1940; repercusiones e interpretaciones.* H. Hermans, D. Papousek, and C. Raffi-Beroud (eds), pp. 101-139.

1997 Privatization of the *ejido* in Mexico. The interplay between legal change and local practice. In: *Natural Resources, Environment, and Legal Pluralism.* René Kuppe, Franz von Benda Beckman, Richard Potz, Keebet von Benda Beckman and André Hoekema (eds.). The Hague, The Netherlands: Kluwer Law International, pp. 72-105.

1995 Changing legislation and a new agrarian bureaucracy: Shifting arenas of negotiation. In *Rural transformations seen from below: Regional and local perspectives from Western Mexico.* Pieter de Vries and Sergio Zendejas (eds.). San Diego, CA: Center for US-Mexican Studies, pp. 49-68.

1992 Local organizations as organizing practices: Rethinking rural institutions. In: *Battlefields of Knowledge: The Interlocking of Theory and Practice in Social Research and Development.* Norman Long and Ann Long (eds.). London: Routledge.

www.ingramcontent.com/pod-product-compliance
Lightning Source LLC
Chambersburg PA
CBHW061210240926
51846CB00006B/1176
* 9 7 8 9 0 8 6 8 6 3 5 6 3 *